Tattoo, Torture, Mutilation, and Adornment

SUNY Series, the Body in Culture, History, and Religion
Howard Eilberg-Schwartz, Editor

Tattoo, Torture, Mutilation, and Adornment

The Denaturalization of the Body in Culture and Text

Frances E. Mascia-Lees
and
Patricia Sharpe,
Editors

STATE UNIVERSITY OF NEW YORK PRESS

"The Greatest Story (N)Ever Told: The Spectacle of Recantation," by Helena Michie, reprinted from *Genders*, no. 9, Fall 1990. By permission of the author and the University of Texas Press.

Published by
State University of New York Press, Albany
© 1992 State University of New York

Production by E. Moore
Marketing by Bernadette LaManna

Library of Congress Cataloging-in-Publication Data
Tattoo, torture, mutilation, and adornment : the denaturalization of
    the body in culture and text / Frances E. Mascia-Lees and Patricia
    Sharpe, editors
        p. cm. — (SUNY series, the body in culture, history, and
    religion)
    Includes bibliographical references.
    ISBN 0-7914-1065-X (alk. paper). — ISBN 0-7914-1066-8 (pbk. :
    alk paper)
    1. Body, Human—Social aspects. 2. Body-marking. 3. Body-marking
    in literature. I. Mascia-Lees, Frances E., 1953– II. Sharpe,
    Patricia, 1943– . III. Series.
    GT495.T38   1992
    391'.65—dc20   91-21296
                                                                  CIP

10 9 8 7 6 5 4 3 2

For our fathers
Anthony C. Mascia
and
Marshall Jamison

Contents

FRANCES E. MASCIA-LEES
PATRICIA SHARPE

1

Introduction: Soft-Tissue Modification and the Horror Within

This volume is about the body as a site of adornment, manipulation, and mutilation, practices with roots reaching far back in the human record, at least 30,000 years.[1] At archaeological sites in Africa, for example, scientists have uncovered bits of clothing on some of our human ancestors; objects from a wide variety of cultures have displayed and recorded forms of body modification for centuries. Bound feet, flesh permanently marked either by a knife or tattoo needle, elongated ear lobes, stretched necks, deformed skulls, shrunken heads—these are practices that have long fascinated the West where they have been viewed as exotic distortions of the body, as is suggested in the standard terminology of "mutilation" and "deformation" itself. Today, as theorists struggle for critical understanding of the West, of its relations of domination and the ideologies which support and mask them, these practices are newly interesting; they no longer serve as bizarre extremes against which

1

we construct our naturalness, but rather to denaturalize the Western body. They are talismans which help us recognize how the body is always culturally constructed. But, of course, even the concept of denaturalization implies a residual belief in the existence of a natural body it seeks to deconstruct, a body outside of culture, a physical norm that grounds human commonality in the face of vast "surface" or cultural differences.

A large mural in the Smithsonian's National Museum of Natural History in Washington, D.C. entitled "Soft-Tissue Modification" testifies to the Western fascination with "exotic" body modifications. The mural is displayed on a center wall in Hall 25, the physical anthropology room. A reading of this room, and the current controversy over it, reveals the tensions that characterize contemporary theorizing of the body, tensions explored and analyzed in the papers in this volume.

Thousands of viewers annually are exposed to this mural: a jumble of superimposed human forms depicting a wide range of body modifications from mostly non-Western contexts (see cover). Figures displaying such unfamiliar practices as head deformation, Japanese tattooing, African scarification, Chinese foot-binding and Mesoamerican tooth filing—as well as such unusual adornments as Burmese brass neck rings, Sara lip plates, South American cheek plugs, and New Guinea nose rings—are contrasted with the foregrounded depiction of a white Western man being tattooed, suggesting a British sailor like those described by Conrad. This juxtaposing of the foreign with the more familiar is a common anthropological technique designed to disrupt the viewer's notion of these practices as exotic. We are invited to overlook the vast differences of culture that separate human beings and find unity in the body: it is portrayed as a ground on which all cultures inscribe significant meaning.

But this anthropological reading is, of course, only one interpretation among many. My thirteen year old son's reaction to the mural was revulsion and horror, despite my painstaking efforts to point out to him the standard anthropological message of cultural relativism that I saw it demonstrating. To him, the practices shown were mortifications, violations to the integrity of the body. The intense pain expressed in the body and face of the young boy apparently during ritual scarification may well have contributed to this impression. The depiction of the sailor's body actually undergoing tattoo invites the Western observer to identify and underscores that the other bodies portrayed have been modified by processes that could also be put to work on the viewer.

My son's interpretation of the mural as a depiction of bizarre and unsettling practices was informed and reinforced by a number of other displays which seem to help construct this particular reading. For while most of the displays in Hall 25 exhibit physical anthropology's contemporary concern with "reading bones" to reveal disease, growth patterns, or the work of evolutionary processes, those nearest the mural seem misplaced by today's anthropological standards in their highlighting of the freakish and the exotic.[2] What, for example, are we to make of the curious Ripley's-believe-it-or-not-like exhibition of the man with the seventeen and a half foot beard? Or "Soap Man," the corpse that decomposed into soap due to a peculiar eighteenth-century burial practice?

But whether one reads in the mural a disruption of ethnocentric attitudes or a reinforcing of the notion of the "other" as exotic or horrific, the underlying message seems surprisingly the same: that the unadorned, unmodified body is an unspoiled, pure surface on which culture works. Either way, this message dehistoricizes and decontextualizes the body. It ignores the particular meaning that both the body and the specific modifications to which it is subjected have for the people being represented. It resolves all bodies into the Western notion of the body as prior to culture and, thus, as natural.[3]

Contemporary theorizing, whether feminist, postmodernist, or anthropological, has contributed recently to exposing "the natural" as a Western cultural construct, calling into question the often taken for granted dichotomy between nature and culture and the ways in which this distinction has acted to reinforce relations of power and domination. Indeed, the body has become an important site for rethinking such binary oppositions as masculinity and femininity, gender and sex, the public and the private, and the cultural and the natural. Contemporary attempts to expose these categories as ideological constructions buttressing Western and/or male supremacy and to disrupt them have focused on the body. Understanding the body not as simple materiality, but rather as constituted within language as in much contemporary thought, is intended to question traditional notions of the body as prior to, or outside of, culture. This move is, of course, just one of the latest attempts in the West to grapple with the relationship between the natural and the cultural and to put the body, and representations of it, in service to this struggle.

A reading of other displays in Hall 25 can expose earlier attempts of using the body to conceptualize the relationship of nature and culture, attempts that remind us of the politics of entangling biology and culture that contemporary biological anthropology has worked to correct. Behind the mural, in the final section of the Hall, is a large display that tells a particular story of human difference fashionable in the West until well into this century. It categorizes physical variations into a racial typology overlaid on a map of the world, superimposing human figures of various colors and shapes onto the continents to show the origin of "the races." By de-emphasizing the wide range of differences within a group in order to highlight differences among groups, such racial classifications reduce arbitrarily the variation in human physical traits into a few simple, static, and discrete categories: Negroid, Caucasoid, Mongoloid.[4] These types, then plotted onto geographic and cultural areas, are made to imply a congruence between physical features such as skin color or cranial size and cultural traits such as kinship terminologies, dress styles, or even sleeping in the nude (see, for example, Morgan 1870:274). This suggests inherent racial identities once seen as part of an evolutionary hierarchy that equated the black with the savage, the white with the civilized. In the heyday of this approach, race and culture were presented as inextricably linked in such a way that a wide range of cultural and behavioral traits were understood as biologically determined.

It is not surprising then that Hall 25 has been the focus recently of criticism from a number of sectors. The allegations of racism and sexism that have been raised, however, reflect more than merely a change in the composition or sensibilities of the viewing public since the construction of the Hall in the early 1950s. The charge by groups whose cultural and physical remains are displayed that they have been misrepresented underscores the current understanding of the museum as an institutional site for depicting not the truth of humankind, but representations of particular cultural notions about identity, human behavior, human similarities and differences, and the role that the body plays in these ideas.

That the body is an important symbolic site in this contest over the representation of cultural identity is perhaps best illustrated by the intense controversy in recent years over the proprietorship and use of Native American skeletal remains in museum exhibitions. Indeed, a number of displays in the physical anthropology room have been modified over the last several years in response to criticism by Native American groups. Thus, the museum, and the representation and display of the body within it, has become an important location where the contestation over cultural identity is being waged in American society.

Like Native Americans, members of other disenfranchised groups have demanded the right to speak for themselves, to tell their own stories, protesting their appropriation into the discourse of the West. What, then, is the responsibility of the Western critic today? Recognizing that representations of the primitive, like those in Hall 25, function to construct the West as normal, as unseen, in contrast with the spectacle of the exotic, critics today struggle to foreground in the images and narratives told about the other, underlying stories about ourselves. The papers in this volume focus on deconstructing some of these Western assumptions, paying particular attention to the construction of femininity and its relationship to the natural and the primitive.

This effort to understand the body as a site for cultural power struggles, initiated by Michel Foucault, has recently emerged as a focal point of research in the humanities and social sciences. Yet this new and intense focus may reflect more than just the new opportunities opened up for research by Foucault's paradigm. Following Lévi-Strauss, Emily Martin suggests that this fascination may express our sense of the demise of the body as we have known it in the West: it may create "the illusion of something which no longer exists but should exist" (Lévi-Strauss 38). Thus, current interest in the body might be understood as nostalgic longing for the simple and knowable in a world in which scientific and medical advances have broken down traditional boundaries. Outside and inside can no longer be distinguished when the technological sophistication of immunological medicine has opened up to view an inner body frequently likened to the vast mysteriousness of outer space (Martin); nature and culture are confused as artificial insemination, pacemakers, implanted lenses, face lifts and sex change operations construct the body as manipulated cyborg (see Haraway). Thus, current preoccupation with deconstructing the body as it has been known may reflect theoreticians' scrambling to understand changes which have already modified the tissue of the body.

In this context, the very question of what the body is is up for grabs, and the contest over the right to define the body's meaning has high stakes. The papers in this volume participate in this debate, using "the body" as a site to ground power struggles over meaning. Like practices of adornment, disfigurement, and manipulation, these papers denaturalize the body, calling into question constructions many in the West have tended to think of as natural. Thus, these papers keep alive a dual perspective on the body. Like the mural in the National Museum, they portray the body as a site of cultural manipulation and focus on the ways in which people are subjugated through the distortion, manipulation, and disfigurement of their bodies or on how such oppression is resisted and protested through the body. They also go beyond such analyses to challenge traditional assumptions about the body, such as those in the work of Aristotle, Freud, or Hawthorne, as well as some newer ones embedded in contemporary scientific discourses, popular culture, and postmodern theories. Informed by an awareness of the politics of representation, each of these papers focuses on how messages about the body are encoded and reinforced in narrative whether newspaper accounts, horror films, stories of the disappearances in Argentina, perfume advertisements, or nineteenth- and twentieth-century fiction.

Thus, in her discussion of popular accounts of the recantation of an accusation of rape in the Webb case, Helena Michie calls into question the idea of the female body as inherently desiring. She exposes the complex workings of the master narrative about women and sexuality in our society. Similarly, Louise Krasniewicz upsets our assumption that reactions to horror films are merely instinctual, visceral responses to violence. She demonstrates the way in which the body, in these films, is used to mediate tensions between the individual and the social. Her reading of the movie *Halloween* suggests that horror films encourage a recognition of incest and bestiality as threats to appropriate sexual relations and the social structures dependent on them. Like Michie, Colleen Ballerino Cohen reads multiple, conflicting discourses that address women in terms of their body and sexuality. Examining various perfume advertisements, women's feelings about scents, and the story one designer tells of creating a particular one, she exposes how the perfume industry promises women autonomy even as it reinforces traditional notions of heterosexual romance and instinctive attraction. Each of these papers interrogates the way in which Western narratives construct the body as true and natural, even as the definition of the natural changes historically.

In her study of female body builders, Anne Bolin exposes the tension between the contemporary ideology of fitness and musculature and the traditional Western ideal of the feminine body as soft and curvaceous. This feminine ideal persists, she shows, even in a sport dedicated to the building of a muscular physique. Her analysis of the conflicting discourses of weight lifting and female beauty highlights the paradox women body builders face in seeking to affirm both their strength and their femininity. Robyn R. Warhol's paper also affirms femininity in the face of an aesthetic standard, one which dismisses female emotion as embarrassingly sentimental. She uses nineteenth-century views of the relationship between the body and emotion—the idea that certain bodily poses directly produce the physical response of tears and corresponding feelings— to dislodge the conviction that high tragedy taps into "naturally" authentic feelings in the audience.

Ñacuñán Sáez's paper, too, is concerned with the relationship among emotions, the body, and language. Examining the record of the Argentine generals, he challenges the rationale underlying torture, that exerting pressure on the body will produce truth. Like Warhol's, his paper is concerned with the conceptual splitting of mind and body and its implication for class relations. Sáez's analysis explores the relationship of "First World" theories to "Third World" bodies, concluding that the exclusive focus on textuality of much recent theorizing may perpetuate this split. Such a double view, focusing at once on a bodily practice and on theoretical symbolism, is also central to our analysis of tattoo. We use the practice of tattooing as an image for contemporary theories that conceive culture or history as writing on the body. We juxtapose the popular film *Tattoo* with Hawthorne's "The Birthmark" to expose how the impulses to mark women as well as to erase the mark of difference have similar consequences for women in Western culture: both attempt to repress a multiplicity of meanings for the body in order to make the female body signify just one thing. Like most of the other contributors to this volume, we highlight the body as gendered and explore the specific ways in which female bodies are conceived, addressed, and erased in contemporary discourses.

The Smithsonian mural displays a similar erasure; it excludes the white Western woman. Although it was presumably constructed in the early 1950s when women's pancake make-up, bright red fingernail polish and lipstick, tight girdles, uplift bras, and stiff permanented hair-dos might well have been classed as "soft-tissue modification," these phenomena go unrepresented. Despite the contortions these women's bodies went through to conform to a narrow ideal of beauty, these practices were constructed as natural and normal and went unseen. The Westerner focused on in the mural is the tattooed sailor who would, by contrast, be seen as a renegade, an adventurer. Somehow his thoughtful, outward gaze and his oversized portrayal suggest that all the other cultural variants depicted take place in his memories of his travels, in imagination. Thus, the mural, like Conrad's *Heart of Darkness*, equates the exotic other with the unspeakable within the white Western male unconscious, while the white Western woman—in Conrad's story the virtuous Intended—represents a purity which demands repression and denial. She can only be constructed as the epitome of civilization, as "whited sepulchre," by constructing an alternative, the uncivilized as the site of "the horror" where all that is repressed has free play. Thus, the story and the mural exempt the white woman, and her body modifications, from scrutiny in order to construct the exotic and thereby reaffirm her "naturalness." These papers, in deconstructing and disentangling the theoretical Western male mind and the body of the other, and in insisting on bringing the white woman into the picture, hope to offer a new mural of soft-tissue modification, one that disrupts the dominant story and enables us to see its workings.

NOTES

1. We wish to acknowledge all those at the Smithsonian Institution who generously gave their time, especially Kathleen Gordon and Felicia Pickering.

2. This interpretation is underscored by the reaction to the room by other anthropologists with whom we have talked about it. Many feel that the room is terribly outdated while others are embarrassed by it.

3. See Marilyn Strathern's "Between a Melanesianist and a Deconstructive Feminist," for a detailed analysis of a different conceptualization of the body. She suggests that notions about the body among the North Mekeo of the Central Province of Papua New Guinea, "do not allow . . . the idea that things are done 'to' the body." For them, "the body is neither subject nor object" (61).

4. Interestingly, Jacques Derrida's work on the idea of *différance*, which focuses our attention on the differences within, is similar to the statistical idea of comparing within-group difference to between- (or among-) group difference. This technique is at the basis of biological anthropology's debunking of the notion of human races as discrete, stable entities since it has been shown that there is frequently more variation within a group on the trait being considered as the basis of classification (for example, skin color) than between this group and another classified as distinct from it. Derrida exposes this process of highlighting differences between groups while obscuring those within as a strategy for promoting the idea of stable identities or concepts. See any recent biological anthropology

textbook for a more detailed description of the problems with the concept of race from current statistical and scientific perspectives.

WORKS CITED

Bartky, Sandra Lee. "Foucault, Femininity, and the Modernization of Patriarchal Power." *Feminism and Foucault: Reflections on Resistance.* Eds. Irene Diamond and Lee Quimby. Boston: Northeastern University Press, 1988. 61–86.

Bordo, Susan. "Anorexia Nervosa: Psychopathology and the Crystallization of Culture." *Feminism and Foucault: Reflections on Resistance.* Eds. Irene Diamond and Lee Quimby. Boston: Northeastern University Press, 1988. 87–118.

———. "Reading the Slender Body." *Body Politics: Women and the Discourses of Science.* Eds. Mary Jacobus, Evelyn Fox Keller, Sally Shuttleworth. New York: Routledge, 1990. 83–112.

Haraway, Donna. "A Manifesto for Cyborgs: Science, Technology, and Socialist Feminism in the 1980s." *Socialist Review* 80 (1986): 65–107.

Lévi-Strauss, Claude. *Tristes Tropiques.* New York: Anthenaum, 1975.

Martin, Emily. "The End of the Body?" Keynote address at the Annual Meeting of the American Ethnological Society. Atlanta, Georgia, April, 1990.

Strathern, Marilyn. "Between a Melanesianist and a Deconstructive Feminist." *Australian Feminist Studies* 10 (Summer 1981): 49–69.

2

The Greatest Story (N)Ever Told: The Spectacle of Recantation

WEBB

In April of 1985, Cathleen Crowell Webb, a born-again Christian housewife from rural New Hampshire, was asked by *People* magazine to tell the exclusive story of her life. Her story appeared among articles on Madonna, Harry Anderson of "Night Court," and celebrity chef Paul Prudhomme; her picture on the cover of the April 29 issue was flanked by smaller portraits of Linda Evans, Phil Donahue, and Tina Turner. Webb, like those who shared the cover of *People* with her, had crossed the line into public spectacle. Her life, her body, and her experiences, like theirs, had assumed the status of "news," that combination of picture and story which signals simultaneously the end to, and the public amplification and appropriation of, private life. Scattered moments from her past—her visits as a young child to her mother in a mental institution, the smell of the

10

eggs her father would cook for her every day—had come together to produce a public account of what she was and had been. The boxed cover picture, showing Webb with her eyes slightly downcast, her hair neatly almost childishly framing a sweet round face, simultaneously alluded to and announced her difference from a wildly coiffed Tina Turner posing with her tongue stuck out in a similarly boxed picture on the other side of the page. On this particular week Webb was bigger news than Tina Turner; Webb occupied the center of the page, center stage; her picture was more than four times the size of Turner's.

Webb's face and story were not introduced into the spectacle of culture that is *People* magazine because she was a movie star, a rock musician, or a talk show host; neither was she gazed at, read about, and listened to, because she was raped. When she had told the story of her rape six years ago, it had received no press; even her hometown paper had not thought it to be good story material. She had made it onto the pages and cover of *People,* become a spectacle, because, six years after the man she had charged with rape had been convicted, she had recanted her testimony, and claimed that she had "made the whole thing up."

The "story" that is Cathleen Webb—made up of so many stories, those she told and those that are told about her—is interesting and problematic for a number of reasons. It raises provocative questions about the law (what is the legal status of recanted testimony?), about sexual politics (would this incident, as many feminists feared, contribute to the cultural assumption that women routinely lie in claiming to have been raped?), and about religion (what was the role of her conversion to evangelical Christianity in her recantation? Did her religion function as an inspiration to be honest or as an inducement to conceal what she might now consider to be the moral taint of rape?). It also raises a more difficult question: what was the truth of the matter? Which of Webb's stories was after all the "true" version of the events of the night in question?

All these questions are important, all of them painful to ask and to begin to answer. The questions I want to ask of Webb's story avoid, in certain ways, and for specific purposes, the issue of whether she was, on the night in question, raped by Dobson or not. Choosing to look elsewhere in her story than at that night, elsewhere in the debate over Webb's recantation than the truth or falsehood of her claims and counter-claims, is itself a painful move for someone like myself who has worked for many years as a crisis counselor for abused women and rape survivors, for someone whose job it was and is, to believe and to assert the daily truth of violence against women. In some ways it would be simpler to write an article defending the truth of Webb's original statement, or perhaps one that moves as Jacqueline Rose's defense of Freud's abandonment of the seduction theory does, to a broader assertion that women who report sexual abuse are testifying, whether the rape literally occurred or not, to the fact and feeling of having been raped by a culture which constitutes itself by the abuse of women. I would like to write both of these articles, because, I believe, whatever this might mean, that they would be true.

The questions I ask here of the story of Cathy Webb are, however, part of a different set of interests, and have to do precisely with the issue of what makes some stories *into* "stories," or more specifically into what I am calling spectacle. I use the term "spectacle" in conjunction with story here because the term suggests the interdependence of narrative and visual elements. The term also outlines a place for the watcher, the audience, the voyeur in a way that story and narrative do not; spectacle also introduces into the picture the gender dynamics of the gaze about which feminists have so compellingly written; although both men and women can be turned into spectacle, the position of gazer is aligned with the male eye and "I," and the position of the gazed-at with the female subject/object. In the sense in which I am using it, spectacle translates the traditionally private into the public by amplifying and rendering visual the interiority of the body and its experiences. Narrative frames and contextualizes spectacle, giving it a meaning, an order, and a teleology. Together, narrative and spectacle produce a story with pictures, embodied on the level of popular culture by glossy magazines and, as we shall see later in my discussion of psychoanalysis, on the level of "high" culture by academic and scientific discourses about the secrets of the body.

To say that Cathleen Webb became a story and spectacle for two weeks in April of 1985 is to say that the American public was finally interested in the story of her life, in her body, and in what happened to it. "Nightline," "Today," and "Good Morning America," *People,* and countless newspapers and television shows across the country literally made space for Webb's autobiography: not merely, as we shall see later, for the story of the night of the alleged and subsequently unalleged rape, but for the larger story of her unhappy childhood, her relation to family, friends, and boyfriends, and her subsequent marriage and conversion. The difference between the two stories—rape and recantation—produced for Webb what could be called a narrative space, a place where Webb could tell a story which was identical with neither rape nor recantation, a place where her autobiography could be made public and her body introduced to the public gaze as a series of pictures, larger in some sense than her life.

If we locate narrative, autobiography, and spectacle between these two conflicting stories, we must, I think, ask why two conflicting stories about rape should produce so fertile a public space. After all, almost every criminal allegation produces its own series of conflicting stories: those told by the defendant, the counsel, and the witnesses for the defense, as well perhaps as those told to the press and those that the various participants in the case tell themselves and their friends. What makes conflicting stories of rape different from those that accrue around other crimes is the structuring framework of what I will call a cultural master-narrative that says that in all cases of rape women are complicitous: that rape is not a rape in the first place, but something else. The story that women lie about rape, whether we choose to believe it or not, informs all stories *about* rape. In undoing the individual story of her rape, Webb was telling this master-narrative; the recantation, then, was not the untelling of a story but the retelling of one infinitely more powerful, a story whose power could only be celebrated publicly and spectacularly. This paper will move beyond Webb and her brief appearance in *People* magazine to look closely at two more authoritative tellings of the master-narrative of rape, two tellings and then untellings of the act of rape: Freud's official fluctuations about the "truth" of childhood seductions that crystallized as his abandonment of the seduction theory and became psychoanalytic doctrine as the theory of infantile sexuality, and Jeffrey Masson's simultaneously academic and popular attack on Freud over the issue of that abandonment. Each position, each story, in its different key, makes a separate but linked series of negotiations with the master-narrative of rape and with its political implications. I use Webb's story, her stories, as, simultaneously, a challenge to and an embodiment of the master-narrative of rape—it is tempting here to call it a mistress-narrative—which, by its lack of authority, its ephemeral position as popular culture, suggests the fertility of master-narratives and brings them to bear on one scarred and culturally insignificant female body.

Webb's retelling of this master-narrative predictably recast itself into a series of individual media stories. These stories are themselves explicitly about the public telling of stories; it is no accident that the media itself repeatedly referred to the recantation in terms appropriate to public spectacle. *Newsweek* called the Webb story "a legal melodrama" (5/20/85) and "a made-for TV movie" (5/20/85), while *Time* elaborated on and literalized the dramatic analogy:

> The drama seemingly will not end until the final commercial is played on the TV movie that may result from last week's flood of offers for Dobson's [the convicted rapist's] life story. (*Time* 5/27/85)

One of *Newsweek*'s earliest articles on the case betrays an almost nostalgic relation to the conventions of television movies. Working consciously within the conventions of sensationalist plotting, the article opens:

> It appeared to embody all the elements of a classic legal nightmare. A confused teenage girl. An innocent man accused of rape. A conviction based on perjured testimony, and a long prison sentence for the young man. But six years later, the case seemed headed for a happy ending. . . . But last week a circuit judge in a suburban Chicago courtroom refused to provide this made-for-TV-movie ending. (*Newsweek* 4/22/85)

If the spectacular ending the story invites is undercut by the realities of a messy, legalistic world, this deviation from the script can itself be absorbed into spectacle. Webb's story is allowed to continue precisely because it is (not yet) a movie; the story spills over into the next week of *People, Newsweek,* and *Time.*

The questions *Newsweek* raises here seem to have to do with the relation of spectacle, authority, and authorship, with the issue of who gets control over the script. While Webb consented to selling her story to *People* and to be interviewed by other magazines and television shows, it is now clear that the (un)alleged rapist, Gary Dobson, wants his own made-for-TV movie which Webb is so far resisting. We can, perhaps, see in this debate between actors in the drama of recantation a battle for control over the spectacular. We can even read the battle cinematically, as a series of efforts on the part of both Dobson and Webb to erect their own frames around the story and to avoid being framed. The interplay among the series of frames the story offers us—Webb's initial "framing" of Dobson, the frame that surrounds Webb's picture on the cover of *People,* the effort of the media to erect their own frame around the story: for example,

Phyllis George's infamous request that Webb and Dobson signal their forgiveness of each other by hugging on national television—forces us as spectators to ask who controls the story, and who, finally, is framed by and within it.[1]

Again, these issues offer us a detour around the question of which (or whose) story is true, a question that pales in comparison to the story of Webb's story-telling, of her two specular and spectacular attempts to control her image and its circulation in the world around her. *Newsweek* raises the issue of control in the initial story of rape:

> Webb, now a married mother of two living in New Hampshire, testified that she had fabricated the rape—ripping her own clothing and inflicting bruises, scratches and cuts on herself—because she thought she might be pregnant. (*Newsweek* 4/15/85)

If we believe the recantation, both the "fabrication" and the ripping of the fabric of her clothing become ways of controlling the access of the world to her body and its experiences. The self-inflicted scratches, which were said to form the letters (words?) "HEHE" and "LOV" on her torso, become a sinister form of self-expression, painful inscriptions on the body oddly reminiscent of some forms of feminist body art.[2]

If we refuse to believe the recantation and believe instead that Webb's original story was true, we can decide, with a number of cynics, that Webb recanted out of a desire for publicity. We might, more sympathetically and more problematically, conclude that the recantation, as a specular inversion of the original rape story, allowed Webb to tell the *original* story of the rape even in the act of denying it—this time to a wide and interested audience. It is at this point, perhaps, that we might look for the source—or at least another location—of Webb's narrative impulse in the story of her conversion; theories of conversion suggest that the identification of a psychological turning point provides narrative coherence and teleology to a life that might otherwise seem meaningless and unstructured. If Webb was looking for a narrative structure for her life, what better turning points than conversion and recantation, what better audience than America?

In this context the "full, heart-rending story" that Webb tells in *People* is especially moving. The story, which the headnote informs us was told "in her own words," moves quickly away from the night of the rape/not-rape to the beginnings of Webb's personal history:

> This is *how* I faked the rape. If you want to know *why* I did it, that is a much longer story. That is a story that goes way back to my childhood. (*People* 4/29/85, p. 36)

The recantation provides a space for the story of Webb's childhood; the rhetorical "if you want to know" hints at Webb's power as an author; she knows we want to know, want to hear why, want suddenly to hear a sad and all-too-common story that is also uniquely her own.

The space that suddenly opens up for the story of Webb's life— and for a series of photographs of her face and body—is structured by the conflict of competing narratives about rape. Read back on the master-narrative which recasts rape as non-rape, Webb's two narratives do not merely cancel each other out; together they make her into a story, and that story is the master-narrative of rape. Since at issue in any rape is the woman's control of access to her body, it is fitting that Webb's storytelling should thematize public access to her body and history. It is especially interesting that while one of the problems of the current law and legal practice around cases of rape is that a woman can be forced to answer questions about her (sexual) past, Webb uses the media to *allow* her to tell her story. The issue of control is, of course, an extremely vexed one; I am not suggesting by any means that Webb broke through the frame of rape law, the mass media, or public opinion; instead, I see her as a figure willing to compromise with the space opened up for her by the master-narrative of rape; consciously or not, Webb used the master-narrative so oppressive to most women, to tell her own story in its margins. The celebration, in the form of media spectacle, of Webb's concession to that narrative, gave her some room to maneuver, her stories—all of them—some room to exist.

The politics of recantation of rape are the politics of spectacle; to tell the master-narrative of rape is to tell a public story, perhaps *the* public story of women and their sexuality. The master-narrative claims that rape does not exist; our culture celebrates the nonexistence, the absence, of rape with a pageant of presence, with spectacles of the body not-violated, with long stories about stories retracted, erased, untold. *People* offers us Webb's body as evidence of non-rape, as evidence of the intactness of the master-narrative. The body that was written upon and scarred in the alleged act of rape is now written about and made public as a guarantee that rape has not happened. The public display of Webb's body mimics the rape that is supposed never to have happened: writing and reading replace writing and raping as we are assured by reading that there was no rape.

The act of recantation brings the news that there is no news; the untelling of news produces stories and pictures, narrative and spectacle. The real story is that there is no story; the spectacularized female body serves as a guarantor of the story that is not one. The story of non-rape also allows us to gaze with impunity at the female body, to mimic, in more benign or at least more sanctioned form, the act of appropriation that is part of any act of rape. Spectacle—the public, published female body—simultaneously assures us that there was no rape and that we may read and gaze at the female body without guilt. I am trying here not—in any simple way—to equate the act of gazing and raping, but instead to articulate the permission to gaze offered to us by the master-narrative of rape, the narrative that assures us that "nothing" has happened.

FREUD

If the spectacle of Webb's recantation seems familiar, it might be because it is structurally so similar to another series of recanted stories of rape: the stories of the hysterics treated by Freud in his development of his theory of infantile sexuality. Like the Webb story, the narrative of the abandonment of the seduction theory upon which Freud's analysis of hysteria is based has to do with the recantation and with the public display of the female body which has (reassuringly) not been raped. Until September, 1897, to invoke the familiar story, Freud believed that the stories his (usually female) patients told him about being sexually abused as children were true, and that the etiology of hysteria lay in what he called the seduction of children, usually by adults or older children within the family. According to Freud's correspondence with Fliess, Freud began, in the fall of 1897, to abandon his seduction theory and to suspect his patients' stories of being untrue; over the next few years he was to develop, in the place of seduction theory, the theory of infantile sexuality, in which his patients' accounts of sexual abuse were re-read as fantasies and recast as examples of infantile sexual energy. Most accounts of the history of psychoanalysis, including those of Freud and many of his most virulent critics, locate the birth of psychoanalysis, as we know it today, in the abandonment of the seduction theory and the discovery of the sexual life of the child.

The recantation of these stories of rape differs structurally from Webb's in one important way; while Webb (if we believe her second story) recanted on her own, Freud recanted for others. We would expect, then, a different relationship between autobiography and recantation in the case of Freud's patients. Critics of Freud's abandonment of the seduction theory have in fact claimed that Freud's new theory effectively silenced his patients, allowing no analytical or narrative space for their autobiography. In abandoning the seduction theory, critics might argue, Freud imposed his biography of his patients upon their autobiographies, rendering their own account of their experiences invalid, unauthoritative. Certainly Freud created an official psychoanalytic narrative, eerily reminiscent of the master-narrative of rape I described above, which acted as a generic biography into which individual stories could—indeed must—be absorbed.

If the abandonment of the seduction theory, this vicarious recantation, effectively silenced the stories of Freud's patients' lives and excised their bodily experiences from the discourse of the new science of psychoanalysis, it would seem that Freud's historically specific reconstruction of the master-narrative of rape produced silence rather than story, blankness rather than spectacle. This is certainly the reading of many critics of the seduction theory, including many who write from a specifically feminist point of view.

I would argue slightly differently and along a historical continuum of explanations of and reactions to seduction theory and its abandonment. First, I would argue, the abandonment of the seduction theory did in fact produce spectacle of the most vivid and public kind: the improbable, shocking, and titillating story of psychoanalysis itself. Perhaps more importantly, the spectacle of psychoanalysis, like the spectacle of Webb's body in *People*, was linked explicitly to autobiography: in this case the story of Freud's own life. Finally, the critical reactions to seduction theory, most notably that of Jeffrey Moussaieff Masson, have themselves assumed the shape and the idiom of public spectacle, and have concerned themselves explicitly with the issue of public autobiography: in Masson's exemplary case with Freud's, with Freud's patients', and finally and perhaps most vividly, with his own. In the rest of this paper I will attempt to trace the relationship between autobiography and spectacle as it appears at certain key moments in the production and popularization of the renunciation of the seduction theory; these moments inevitably become embodied in the stories of the three people or groups of people mentioned above: Freud himself, his hysterical patients, and Jeffrey Masson.

We can begin to locate the autobiographical principle at work in the first moments of the renunciation in Freud's famous letter to Fliess dated September 21, 1897. In the letter, Freud clearly saw the abandonment, as he tended to see most of his discoveries, in autobiographical terms. Autobiography was for Freud a common idiom in which to produce scientific accounts; not only did he place himself and his own psyche at the center of his scientific scrutiny in seminal texts such as *The Interpretation of Dreams*, but he read the notion of scientific progress back on his development as a scientist, as a thinker, and as a human being. In the crucial letter to Fliess in which he announced for the first time that he was dissatisfied with the seduction theory, he moves quickly to a description of his own affect, his own ontology. He is telling two stories here that become inseparable: the story of psychoanalysis and the life-story of its first practitioner:

> If I were depressed, confused, and exhausted, such doubts [about the validity of the seduction theory] would surely have to be interpreted as signs of weakness. Since I am in the opposite state, I must recognize them as the result of honest and vigorous intellectual work and must be proud that after going so deep I am still capable of such criticism. Can it be that this doubt merely represents an episode in the advance toward further insight? (265)

The rhetorical question that ends this paragraph projects Freud and his readers into a progressivist scientific narrative that depends for its triumphant conclusion on "episodes" of doubt and error. He is asserting a double narrative that depends partially for its persuasiveness on what feminists have come to call the authority of experience; Freud reads his own body, his own moods for symptoms, much in the way he reads his patients'. Ironically, while he trusts his own body to tell him the truth in the face of confusion and contradiction, he begins to decide here that the accounts his patients offer of their own bodily experience is invalid, unscientific, without further translation into the emerging discourse of psychoanalysis.

Freud's remarkable and frequently endearing candor about his self-doubt and his constant need to revise become tropes for his later work. I see them not merely—and perhaps not most importantly—

as self-effacing admissions of error, but as strategies of control and containment. If his self-editing and autobiography become the idiom of psychoanalysis, then psychoanalysis itself becomes the story of his life, and the authority of Freud's own experience becomes the touchstone for scientific truth. His inclusion of past mistakes into the text of, for example, *Three Essays on Sexuality* becomes, then, not so much a humble admission of error, as a tactic that instead establishes the authority and teleology of his own story of science that is almost impossible to refute because it is so evidently the story of his own experience. His autobiographical authority is particularly vital for his explanation of the abandonment of the seduction theory because the issue at stake is precisely the authority of experience. By proclaiming himself to have been, at particular historical moments, in error, he proclaims the authority of the present, embodied in himself as editor, to make a coherent narrative of the past. Since this drive to coherent renderings of the past is both the structure and the content of his work, it is not surprising that it should be thematized so frequently in his writings.

I would suggest, however, that the issue is not simply one of competing autobiographies, of the silencing of one series of life stories by one which is louder, male, and more authoritative. Clearly at work in Freud's reconsideration of the seduction theory is a profound and problematic identification with all of the actors in the drama of childhood seduction. Marianne Krull has looked to Freud's biography, and particularly to his relation with his father, in an effort to explain some of the problems with the abandonment of the seduction theory. Beginning with Freud's objection in the letter to Fliess that "In every case *not excluding my own*, the father had to be accused of perversion," and working through the dream that Freud had on the night of (or before) his father's funeral, where Freud found himself instructed by a placard to "close an (the) eye(s)," Krull has suggested that after the death of his father, at the very time of his reconstruction of the seduction theory, Freud was following his father's "mandate" not to go back into his own past, not to see that his father, like the fathers of so many of his patients, was a seducer (42, 56; italics in original). Krull's argument is a complex one that can neither be fully presented nor evaluated here. Again, I am not so much concerned with the truth value of the stories told by Freud or his critics, as with the autobiographical and spectacular impulses at work in their construction. Clearly, in Krull's account, Freud, as a father, a son, or both, was identifying with his father and constructing an elaborate scientific narrative to protect his memory. In the process, according to Krull, he was protecting himself from the memory of his own seductions at the hands of a series of adults in his family, seductions which, according to Krull, were hinted at by Freud's mother in her tales of his childhood (59–60).

The issue according to Krull, then, is not merely the silencing of his patients' autobiographies, but the repression of his own. The autobiographical impulse of the Fliess letter situates itself only in Freud's adult life; he has disobeyed his own injunction in "The Aetiology of Hysteria" to trace "determining qualities" of symptoms back into childhood (185–186).

While Krull sees the identification with his father as the psychological locus of Freud's repression of his autobiography, she does not investigate what I find to be a more problematic identification with his patients and, structurally speaking, with daughters. In the same letter to Fliess where Freud resisted the seemingly inevitable assault on the reputation of his father, Freud metaphorized his shame at being wrong about his theory of seduction in strictly feminine terms. In a reference to an old Jewish joke from the "collection" that was later to become *Jokes and Their Relation to the Unconscious*, Freud says cryptically about himself: "Rebekka, you can take off your wedding-gown, you're not a *Kalle* [bride, whore] any longer" (*Letters* 266). The slippage "kalle" allows from "bride" to "whore" is interesting for a number of reasons. Most obviously, it is unclear whether the abandonment of the seduction theory turns him from a bride to a whore or a whore into a bride, whether the sexual taint resides in accusing men of molesting their daughters or in closing one's eyes to the fact. Perhaps more importantly, the oscillation between bride and whore allows Freud only feminized alternatives. Throughout his work and his life, in subjecting himself—and to a certain extent his privacy—to the gaze of others, Freud constructs for himself a feminized position. This should not be taken to mean that Freud, in the process, gives up certain kinds of powers traditionally associated with men, but rather that the power dynamics of spectacle and autobiography provide a subtext to the master-narrative that depends on the male prerogative of interpretation and diagnosis. Freud's "feminine" self-construction is simultaneously supported and undercut by the context of his admittedly erotic relation to Fliess; his identification as bride scripts him as the erotic other while his position as father of psychoanalysis allows for and indeed produces a homosocial discourse in which the letter—and the relationship with Fliess—are embedded.[3]

Krull's reading of the letter to Fliess, in pointing out only that

Rebekka was the name of his father's mysterious second wife, and therefore, presumably, a symbol for the mysterious power of his father's sexuality, ignores the narrative within the master-narrative that positions Freud as daughter as well as father. By impersonating the bride turned whore by the joke and the pun, Freud is expressing not only the perversion of the father, but the sexual taint of the daughter who is no longer the innocent victim of rape, but rather an already formed sexual subject with desires of her own. Freud's identification with the daughter can, of course, be read as a proto-feminist gesture or simply (?) as a further and more insidious appropriation of the female body. By constructing for himself a feminized position, Freud bolsters the authority he has already granted himself to speak for women.

Freud tells the story of female sexual shame in the context of a joke that, in accordance with Freudian readings, both represses and reveals his own autobiography, transforms and expresses his own gender identity. The series of autobiographical gestures he makes and then undoes are oddly reminiscent of Webb's. His recantation, like Webb's, allows him to tell his life story—and, later, the life story of psychoanalysis—as he wants it to be told.

Like Webb's religious conversion, Freud's conversion to the theory of infantile sexuality can be read, by him and by others, as a turning point that gives meaning and shape to earlier, seemingly unconnected fumblings. In his *Autobiographical Study*, published twenty-eight years after the letter to Fliess, Freud placed the abandonment of the seduction theory at the center of what he was later to refer to as "the story of my life and the history of psycho-analysis . . . intimately interwoven"(71). In the *Study* seduction theory is presented as "an error . . . which might well have had fatal consequences for the whole of my work," his abandonment of it as a sign that he had "pulled [him]self together" after "a severe blow" to his "confidence" (*Study* 34). The pulling together of a self shattered by a breakdown in confidence mirrors the pulling together of the two stories that make up the *Study*: the story of self and science-in-self.

THE FEMALE PATIENT

Freud's teleology of life and science in place, it is difficult to imagine a counter-narrative, a counter-biography offered by one of his patients. While a host of critics have read Dora's famous exit from Freud's office as a gesture of resistance, her silent leave-taking must be recast as narrative for her by others: like Freud, feminist and other critics of his technique must "read" Dora's symptoms and her silences to produce coherent narratives; they ultimately, of course, reveal more about "Dora's" readers than about the person we might provisionally designate as Dora herself. There is no doubt, however, that Dora's story and body have become feminist spectacle, and that Dora's disruption of her analysis has been cast by feminists at least in theatrical terms as a dramatic exit, a breaking of the frame of the analytic scene/session.[4]

Despite the attention paid to "Dora" and to Freud's (mis-) readings of her sexuality, her case has not specifically been used to examine the abandonment of the seduction theory. In constructing his theory of the abandonment, which, like Krull's, is heavily dependent on Freud's biography, Jeffrey Moussaeiff Masson has turned his attention in his controversial first book to Freud's mysterious patient, Emma Eckstein, and then to all women patients under analysis. Masson repeatedly uses the trope of biography to disrupt Freudian theory; like *People* magazine, he is interested in "heartrending accounts"(20), in life stories. In the introduction to his recent collection of nineteenth-century articles on female sexuality, *A Dark Science*, he combats what he feels to be the hegemony of theory with an appeal to the authority of experience:

> Five minutes with a survivor of incest will tell you more about it than fifty years of textbooks on psychiatry. Let me quote from a letter I received recently from a victim of incest. (*DS* 26)

It would seem from the introduction to *A Dark Science*, as well as from Catherine MacKinnon's preface comparing the silencing of women by pornography and by psychiatry, that Masson would be primarily interested in recovering women's voices, in giving us their accounts of their own mental, physical, and sexual lives. Instead, there is a disjunction between the introduction and preface which depend heavily on (usually female) first-person narratives, and the body of the text, which tells only the "official" story of male-authored "scientific" articles.

MacKinnon's preface opens, a little eerily, with a quotation from a little girl overheard (presumably by MacKinnon) at a "suburban Los Angeles shopping mall." She is saying, in Spanish, which MacKinnon translates for us, "You're hurting me, Daddy, you're hurting me." The epigraph establishes the idiom of the prefatory material; it is personal, autobiographical in two senses; it is a story from the daily life of the little girl, but it is also, perhaps more importantly, an episode from MacKinnon's life, since it is she who presumably overhears. The use of an epigraph from "real life," undocumented, unauthorized even by the idiom of the psychiatric "case" or even the confessional letter (which Masson, of course, uses later), immediately foregrounds the issue of belief. We are asked to believe that the little girl did say this, that it was overheard, transcribed, and translated correctly by a reliable witness. We are asked to believe that the relationship of the father to the child was abusive, that the "hurting" was not accidental. We are asked, in effect, to construct a narrative about the child's life and to place this moment within an elaborate speculative biography: all this without scientific "evidence" of any kind. If we believe the little girl, and the overhearing, overseeing ear and eye, if we believe, for instance, that it was MacKinnon who overheard and that we trust MacKinnon, we will be able to read on with a foundational belief in the authority of experience. When MacKinnon goes on to explain how women were routinely disbelieved in court and in the psychiatrist's office, we will have made the leap of faith necessary to believe them in the teeth of science. We will now believe Masson's letters—after all, these can be documented—and will extend our belief to all the victims and "mental patients" who figure in the pages of *A Dark Science*.

By the end of the introduction, however, the victims stop speaking. The rest of *A Dark Science* consists of shockingly misogynist articles written in what Masson at any rate claims are important nineteenth-century medical journals. In the course of the book we read of clitoridectomies, cauterization of the female genitalia, straitjackets to prevent masturbation: a veritable litany of atrocities. Masson's contention is that psychiatry has not radically improved in its sensitivity to women since the days of the straitjacket and the hot iron.

While the articles are both as shocking and interesting as they are obviously meant to be, there is also something odd about the project of the book as a whole, and that oddness seems to have precisely to do with spectacle. The cover drawing, an illustration from "Une Semaine de Bonté" by Max Ernst, depicts a woman asleep in a curtained and beruffled bed, arms flung sensuously over her head to reveal her breasts. As the eye travels from this scene of luxury we see, at the right and back of the picture, part of a dark, male figure with a mustache, half-hidden by the bed curtain, who is surveying the prone female figure through what seem to be prison bars. Even the curtain seems to be darker on the man's side of the picture; the woman's life and health seem to recede before our eyes as our gaze is wonderingly drawn deeper into the picture and into complicity with the watcher. Like the doctor-figure, we are dressed, awake, and watching; like him, perhaps, we take advantage of the body half-revealed to us by sleep.

Masson, or whoever chose the cover picture, clearly meant for it to be arresting, meant it to encapsulate the problematics of male scientific scrutiny of the female body. The place of the reader, however, makes the picture troubling, especially when we turn to MacKinnon's first sentence:

> Reading these texts (the scientific articles) is a lot like reading pornography. You feel you have come upon a secret codebook you were not meant to see but which has both determined and obscured your life. (DS xi)

If these texts are indeed pornographic, what is our role in reading them? The texts and the cover picture both ask of us that we gaze at the naked female body and at the unspeakable operations performed upon it by science. Are we not then complicitous in the spectacle made of the female body?

MacKinnon would seem to try to deflect this question by the use of the generic "you" in the passage quoted above. If you the reader are also you the victim, then you are merely gazing at "your"self, and the story of the female body is your autobiography. The you denies history, collapses almost a hundred years into one act of looking, tries to convince us that nothing is changed and that the women of the picture and of the article are still, somehow, us. This compression of history allows the book to exist; these articles become not so much medical curiosities as descriptions of our own lives—autobiographical moments.

Again, spectacle is justified by the power of autobiography, the entrance of the female body into possibly corrupting and/or painful public space rendered meaningful by the power of a life story. Again, this entrance is effected under the auspices of the master-narrative of rape. Unlike Webb, however, the women simultaneously silenced and brought into textuality by this text did not choose—however problematically—to be brought once again into spectacle. Unlike Webb, they are made into spectacle completely vicariously. There is, of course, another vital difference between the case of Webb and these "cases": Masson and MacKinnon invite the mutilated female body into the circumference of our gaze as a resistance to the master-narrative with which Webb finally seems to comply. Oddly enough, however, it is the master-narrative that has the last word, as *A Dark Science* enacts the contradiction between the rape that was and the rape that never was or could be in the conflict between the prefatory material and the "body" of the text. Once again, the master-narrative makes room for the spectacle of the female body, simultaneously imprisoning it and allowing it a muted and problematic life story.

MASSON

The final autobiography and the final spectacle in this chain are, of course, those of Masson himself. The public knows his face and body well; like Webb's it has been on television and in a number of popular newspapers and magazines. It stares at us from book jackets, it appeared on Public television's hour-long special on Freud and his legacy, and is described erotically by Janet Malcolm, first in two long articles in the *New Yorker*, and then in the book she fashioned from the articles: *In The Freud Archives*. Malcolm opens her discussion of Masson's assault on Freud's recantation with an ambivalent tribute to Masson's ability to "draw a certain perplexed attention to himself"(3). She moves quickly to a consideration of Masson's physical attributes:

> He was good-looking in a boyish, dark, mildly Near Eastern way. (The photographs that were presently to appear in the *Times*, *Newsweek*, and *Time* make him look more exotic than he does in life, and a bit plump and spoiled). (*FA* 3)

and then to an assessment, through Victor Calef, of the figure cut by Masson and his wife as a couple. "They were so beautiful, so brilliant, so delicious" (*FA* 4). Malcolm is obviously both describing and perpetuating the image of Masson as body, as individual; in addition to dwelling on his appearance, she quotes him at some length about his sexual habits and his self-confessed inability to be monogamous. As Masson himself points out in his preface to the Penguin edition of *The Assault on Truth*, Malcolm's articles set the tone for "personal" attacks in more formal reviews of *The Assault on Truth*:

When this book first appeared in hardcover, the many reviewers who wrote about it concentrated their criticism on the
supposed character of the author . . . My character was attacked, my motivation was attacked, but my arguments, and
even more importantly, the new historical material I brought to
light, were not adequately addressed. (xvii)

While Masson's critics accuse him of personal ambition, of "wom-
anizing" on the one hand, and of being entangled with a highly
eroticized homosocial competition with Freud on the other, of making unnecessarily personal attacks on Freud, and of insisting too
much on his own personality and public persona, Masson accuses
his critics of mistakenly focusing on his character and autobiography. Both Masson and his critics work within the idiom of the personal. Like the story of Cathy Webb, the story of Jeffrey Masson left
its referential field for the pages of the popular press. Like Webb's,
Masson's face became public property, his sexual life a household
story.

It is easy to detect in Masson, as it is in Freud, an uneasy identification with women. Since the publication of *Assault*, Masson
has become more overtly feminist, as his association with Catherine
MacKinnon suggests. In the *Nova* special on Freud, Masson represented the *only* critique of his work from a feminist position; wom-
en, more than conspicuously absent from the program, were represented by the familiar face of Jeffrey Masson. Malcolm probably
senses this identification in the opening paragraph of *In the Freud
Archives*, where, in a fairly elaborate extended metaphor she com-
pares other analytical candidates to "shy, plain girls at dances," and
Masson, by implication, to a belle:

(Masson) . . . not only steered clear of wallflowers but was
dancing with some of the most attractive and desirable partners at the ball: with well-known senior analysts, such as Sam-
uel Lipton, of Chicago. (3)

Despite her later depiction of Masson as a playboy, the sexual dynamics at work in this metaphorical "introduction" to Masson at
the dance are worth exploring.

Masson's own positioning of himself as a feminist voice is both
deeply problematic and moving. He claims in his preface to the
Penguin edition of *The Assault on Truth* that the reviewers who

"followed Janet Malcolm's lead in focusing their attention on my personality" were "almost all men" (xx). On the other hand, he admits that "women, sympathetic to the research I have done," have complained that he gave no credit to feminist work on incest. One might imagine that Masson, once in a position of such power and centrality in the psychoanalytic community, is now taking what power he can from a marginal position. Malcolm's dance metaphor, however, hints at something feminine already in place at the height of his power, and suggests that his previous institutional position was itself constituted out of a complex identification with a feminine position which, like Freud's, could—and perhaps must—coexist with traditionally masculine forms of power.

No matter how vexed his identification with women, we must admit that in some complicated and perhaps twisted sense, Masson's autobiography is a site from which the story of women is being told. It may not be the story feminists on either side of the debate over seduction theory would like to hear, but it is, right now, the story, and Masson's face and body are the spectacle of recantation. While some feminists, like Jacqueline Rose, have maintained that the abandonment of the seduction theory allowed Freud to tell the story of all women, whether or not they had actually (literally) been raped, others, like MacKinnon have seen it as a renunciation and a silencing of women's experience. The issue is presently being debated, however, not in feminist journals, but in the popular media, where Masson's body and story, have, oddly enough, become synecdoches for the bodies and stories of women psychiatric patients and rape victims. To address the issue of the seduction theory, one must address oneself to the spectacle that is Jeffrey Masson. Perhaps even more problematically, feminist lawyers and judges, as well as rape victims, must live with and try to understand the specter of Cathleen Webb and women like her, who, in words from outside the lexicon of psychoanalysis, outside this culture's official and authoritative discourse about sexuality, violence, and identity, tell in words that are in some sense their own the master-narrative of rape.

NOTES

1. For a discussion of cinematic framing and the female body see Betsy Erkkila, "Greta Garbo: Sailing Beyond the Frame," *Critical Inquiry*, 11.4 (June 1985): 595–619.

2. See Lucy Lippard, *From the Center: Feminist Essays on Women's Art* (New York: E. P. Dutton, 1981), 121–138.

3. My readers for this article have correctly and provocatively suggested a possible counter-narrative which sees in homosociality itself a source of the impulse towards narrative. This analysis cannot be fully developed in this paper, but it is useful to think of the relations between homosociality and the places it allows for the recounting of female experience.

4. See, for example, Nina Auerbach, "Magi and Maidens: the Romance of the Victorian Freud," *Writing and Sexual Difference*, Elizabeth Abel, ed. (Chicago: University of Chicago Press, 1982), 111–130, and Maria Ramas, "Freud's Dora, Dora's Hysteria" in *In Dora's Case: Freud—Hysteria—Feminism*, Charles Bernheimer and Claire Kahane, eds. (New York: Columbia University Press, 1985), 149–180. Any of the essays in this collection could, of course, be used to trace feminist fascination with Dora.

WORKS CITED

Freud, Sigmund. "The Aetiology of Hysteria." *Standard Edition* III, 1986. 191–221.

———. "Postscript." *An Autobiographical Study, Standard Edition* XX, 1935. 251–258.

Krull, Marianne. *Freud and His Father.* New York: W. W. Norton, 1986.

Malcolm, Janet. *In the Freud Archives.* New York: Random House, 1983.

Masson, Jeffrey Moussaeiff, Ed. and trans. *The Complete Letters of Sigmund Freud to Wilhelm Fleiss.* Cambridge: Harvard University Press, 1985.

———, Comp. *A Dark Science: Women, Sexuality, and Psychiatry in the Nineteenth Century.* New York: Farrar, Strauss and Giroux, 1986.

———. *The Assault on Truth.* New York: Penguin, 1984.

3

Cinematic Gifts: The Moral and Social Exchange of Bodies in Horror Films

Looking for places, events, objects, and texts that speak to the encoding of the body in society, nothing comes more vividly to my mind than contemporary horror films. An amazing array of violations are encountered by human bodies in these narratives—they get chopped, stabbed, crushed, dismembered, exploded, melted, electrocuted, drowned, and shot. These bodies are alternately dropped and levitated, burned and frozen, enlarged and shrunken, possessed and exorcised, or metamorphosed into any number of odd creatures. Horror films seem to point a bony finger at the vulnerable yet powerful body as the focus for our horrible attentions. Because of this preponderance of bodily associations, the horror film can be approached, albeit cautiously, as an important narrative site at which ideas about the social use and abuse of the body are both encoded and enacted.

The manipulations to which the body is subjected in the horror film cannot be approached as either simply realistic or simply imagined. Films in general, and horror films in particular, reside in that border zone between the intangibility of narrative and the concreteness of everyday life. A narrative entertainment film allows an audience to interact with non-ordinary circumstances and persons while also requiring the audience to compare familiar work-a-day narrative structures with the uncanny, marvelous, or fantastic ones of the film. The film becomes a reference point for reality (and so people say, "That's just like in the movies!") at the same time that it is recognized as being "just a movie." The horror film is among the most conventionalized of film genres in part because of this need to have audiences accept these often bizarre horror images as somehow imaginable. The power of classic film narrative is that it creates an experience that simultaneously embodies highly constructed images and images that seem more real than the real itself.

In a more sociological vein, Vivian Sobchack has characterized the horror film genre as being "primarily concerned with the individual in conflict with society or with some extension of him[sic-]self" (30). She proposes that conflict is personalized for the main characters in the horror film and this conflict concerns the inevitable struggle against the internal evil that is said to motivate much of human action. At the same time, Sobchack claims that horror films are more generally about issues of moral chaos and the disruption of the "natural" or divine, god-given order. She distinguishes all these horror film characteristics from those of the science fiction film where the dilemmas are not moral and personal but rather firmly anchored in the external threats to "manmade" society itself.

Although Sobchack rightly insists that horror films focus on individual danger and chaos and that larger social dangers seem to inhabit science fiction films, she also acknowledges that making too much of a distinction between these two realms of conflict is counterproductive. I want to expand this cautiously articulated idea of hers and suggest that while we can and should make distinctions between the moral and the technological, and between the individual and the social, we also have to recognize the deep interdependencies and interactions of these arenas. To say that the horror film is less about larger social issues than it is about personal struggles is to underplay the inevitable conflict between self and society that inhabits all human action and thought. The social contract which humans uneasily embrace when they construct their imaginative communities forces them to be constantly evaluating individual needs and desires against those matters that are imperative for the continuation of the community. Films, then, enact a type of ritual moment in which the social contract is threatened and humans are at risk in their identities as social beings. Stephen Prince richly states this sociocultural element in horror films when he explains that, "The horror film may be regarded as a compulsive symbolic exchange in which members of a social order . . . nervously affirm the importance of their cultural heritage" (28). Horror films are concerned with the social aspects of both individual and group identity when they address "the persistent question of what must be done to remain human" (Prince 28).

As part of a network of socially constructed representations that are designed to engage our interpretations about current community issues, films will necessarily address the fragility of the social contract and of community and individual identity. Horror films can productively be seen as representations of the negotiation between the individual and the social. The movies are a form of social exchange designed not only to display this struggle but also to seal this contract by offering simple solutions to complex social problems. Films act in our society the way traditional representations (such as myths) do in traditional societies—they present guiding narratives that investigate, identify, depict and then neutralize any threats to the social contract. I do not want to stop, however, with the film as merely another text whose telling automatically can be used to promote the dominant culture. As powerful as such narrative treatments may be, there is another element of film that requires careful consideration.

It would not be unreasonable, at least from the social science perspective, to define the movies as a kind of "gift." In his classic analysis of gift exchange in "archaic" societies, French social scientist Marcel Mauss explains that gifts, or "prestations," are universal aspects of human communities, a part of the total interaction of social, economic, political, and personal activities in a society. Gift giving and receiving are not merely the exchange of goods and services (including storytelling), but also the exchange of values, moral judgments, and most significantly, social obligations. Society is built on such social exchanges and a refusal to take part in them is an antisocial act that threatens the very foundation of the social contract.

In the pres(en)tation of a film, the film producer/distributor offers the moviegoer a story whose seeming purpose is to offer entertainment and temporary refuge from the monotony or tensions of everyday social existence. American narrative films, especially those employing generic conventions, usually claim to demand no more than that the audience sit back and lose itself in the story. Mauss, however, warns us about the duplicitous aspect of the gift— every gift has hidden social and moral agenda attached to it that should not be ignored or forgotten. Prestations, he explains, "are in theory voluntary, disinterested and spontaneous, but are in fact obligatory and interested"; the behavior that accompanies the gift is based on "formal pretence and social deception" (Mauss 1).

What "Hollywood" hides in the cinematic gift are the moral narratives needed to sustain the society that produces and supports these stories in the first place. Hollywood packages this social gift in tantalizing visual images, romantic myths, economic rationalizations, emotional justifications, political explanations, and convincing religious and legal value systems. The generic narratives that characterize much of contemporary cinema are intended to be conservative—most gifts are—and are designed to create a sense that the viewer has a moral obligation to heed the warnings of the film. The advice which constitutes the cinematic gift is concerned with how to be identified and how to function as a proper human being in the particular society that offers the film.

The cost of the cinematic gift to the viewer is not merely the price of the entry fee; movie viewing cannot be rationalized in such utilitarian terms. The price one pays for movie viewing is the price of living in communion with others—the viewer is exposed to socially condoned prescriptions about how society and individuals should ideally function. If the film gives us social propaganda couched in terms of an entertaining story, we must, according to the laws of reciprocity, give it back a promise to maintain the type of society that will continue to produce and support such stories. For movie viewers, the price to be paid is membership in a social community that sees life as if it were "just like in the movies," movies that are created and controlled by a small number of people who maintain a limited vision of the social as well as the individual body. This membership is simultaneously comforting (it provides company and simple answers to a lot of questions) and claustrophobic (it inhibits radical and/or individual interpretations), but it is never entirely escapable.

The cinema is not only a gift itself, but being inherently social it also enacts gift-giving rituals within its narratives. One of the primary gift exchanges enacted in the horror film is the exchange of bodies. Since ensuring the movement of all types of gifts through the proper channels is necessary for maintaining social order, the gift of bodily exchanges must be carefully controlled by defining who or what can be exchanged and how this can be done. What is socially acceptable and what is taboo in bodily exchanges help define who is a good or real member of a community and who is not. A taboo is best seen as a description of who can or cannot engage in certain types of behavior; it does not mean a taboo behavior is completely forbidden (Arens 6). Obeying or breaking taboos involves making moral judgements about the proper uses for human bodies in order to establish who is "us" (kin and kind) and who is "them" (non-kin and of another community/species). In other words, taboos are commonly used to establish the social/antisocial as well as the human/animal line, with all "others" being categorized apart from the social body with animals and dangerously antisocial people.

Two of the main body taboos that help define who is social and who is other are those concerning incest and bestiality. Stories about incest and bestiality are potent narrative gifts that offer useful moral fables about the dangers of improper lovemaking. Each society's particular limitations on bestiality, or interspecies lovemaking, and incest, or intrafamilial lovemaking, are good indications of what that society defines as the human/animal, us/them, social/anti-social boundary (Shell 123). In the 1981 horror film *The Howling*, for example, the taboo on bestial sex is challenged by a group of California werewolves. The head of the group (which calls itself the Colony) is a psychologist who has written a book called, conveniently enough, *The Gift*. His gift, like Mauss's, is one that unites people in a community in which they get caught up in a series of reciprocal obligations that include not only sex and companionship, but also blood and victims to feed their metamorphoses into wolves. The psychologist explains that his version of the gift offers the knowledge of the beast or animal within. While denying this animal nature is necessary for nonbestial communities, questioning that humans are intertwined with the animal world would be devastating for his alternative Colony. The gift in *The Howling*, this time unlike Mauss's gift, is designed to blur the line between humans and animals and to generate instead a boundary-crossing world of beastly human pleasures.

I want to explore the theme of the taboo bodily exchanges of incest and bestiality more closely as they are presented in another horror film, *Halloween*. My purpose in this analysis is to show that horror films, instead of just being trashy fodder for the adolescent imagination, are among the most socially loaded obligatory gifts offered by a society which has an enormous stake in keeping bodies, especially developing adolescent ones, exchanged in the proper ways. Nothing breaks down social, moral, and legal boundaries faster than bodies that are thought to be misbehaving, and any set of narratives that frequently and fervently dedicates itself to this topic demands our attention. Audiences never tire of horror films, Prince suggests, because the films speak to the most basic questions of individual and group identity (28). Although horror films (and *Halloween* in particular) have received frequent attention in recent analyses of contemporary film, a rereading of the genre that emphasizes the social rather than the psychological, philosophical, or artistic elements of this genre will add a new dimension to the ways the current enormous body of horror films can be considered.

Halloween was a low-budget 1978 film directed by John Carpenter and produced by Debra Hill (and written by both of them). Both Hill and Carpenter went on to other work in Hollywood as a result of the tremendous commercial and popular success of this movie. *Halloween* is considered one of the first and "best" in a long line of teen slasher/stalker films (a subgenre of the horror film) that continue to be popular to this day. The film begins on Halloween night in 1963 in the small town of Haddonfield, Illinois with the first of a two part temporal generic structure in which a past event is enacted and then used to explain present day events (Dika 93). As the movie begins, we see from the point-of-view of some unknown person looking into the window of a home where a teenage girl and her boyfriend are petting on a couch. The parents of the girl have left and given her the task of babysitting her little brother, Michael. The young couple goes upstairs and from outside we see a bedroom light go off. The voyeur moves into the house and takes a knife out of a kitchen drawer. The camera, continuing the subjective point of view, shows the boyfriend hurrying out the door and making an insincere promise to call the girl later. Upstairs, as the girl is sitting barebreasted at her vanity table combing her hair, the still unidentified voyeur picks up and puts on a Halloween mask that the boyfriend was previously wearing; he then enters the girl's bedroom. He stabs the girl who calls out and identifies him as her brother, Michael. The attacker then quickly walks outside where the returning parents are approaching the house. In the final scene of this past-time sequence, a boy about six years old, Michael Meyers, is shown wearing a clown costume and standing with a bloody knife in his hand, flanked by his surprised and frozen mother and father.

The stalker/slasher film's second temporal segment takes place in the present with the killer returning to "take vengeance on the guilty parties or on their symbolic substitutes" (Dika 94). *Halloween* picks up again fifteen years later on the day before Halloween night when Michael Meyers escapes from the mental institution that has been holding him. He returns to his hometown and conducts a murderous Halloween night rampage. A bookish, sexually innocent, teenage baby-sitter named Laurie catches his attention, and after murdering several of her friends as they engage in sexual activities, Michael pursues her in the house where she is babysitting for two children. Michael Meyers in turn is pursued by his psychiatrist, Dr. Loomis, who is himself obsessed by Michael and who tries to convince the authorities that the killer has returned and is dangerous. At the end of the film, Michael, after cornering and wounding Laurie, is shot by Dr. Loomis but mysteriously disappears from the spot where he had fallen, leaving behind a shocked Dr. Loomis, an hysterical Laurie, and several movie sequels.

In an excellent analysis of this slasher/stalker horror subgenre, Carol Clover decides *Halloween* is about gender-identity issues in which both the pursuer and the pursued are struggling over their sexual orientation and activities. The anxieties about sexuality and the ambiguous gender characteristics that are shared by perpetrator (Michael) and victim (Laurie) make them partners in a plot that tends to question stereotypical gender representations. *Halloween* marked a shift in the horror genre, according to Clover, because it encouraged an identification by both male and female viewer with the masculinized but victimized female protagonist (the Final Girl who is left to battle the killer). But by eventually assigning that experience to an other rather than accepting it as part of themselves, the male audience members can remain safe from the ultimate fate of such an identification (acknowledging the female within) while the female viewer can only value the Final Girl for her masculine characteristics and not for her equally prevalent feminine ones.

Clover's innovative foregrounding of issues of gender is an important advance in the analysis of horror films. However, even more general questions about the differences not just between males and females but between humans and antihumans always haunt the horror film and not always in stereotypically gendered ways. To get an adequate understanding of the ways bodies are used socially in horror films, questions about these other boundary crossings must be asked and addressed. To get to these issues in *Halloween*, we first need to clarify some details of the story. Specifically, we need to know why Michael Meyers kills his sister and not someone else; why he breaks out of the mental institution after fifteen years and not sooner or later; and why he pursues Laurie and kills her friends instead of some other girl and some other group of teens.

Michael's murder of his sister is on the surface the inexplicable act of a mad person. The murder appears to be directly related to Judith Meyers's sexuality—she is murdered while still naked from her illicit sexual encounter with her boyfriend. Yet to simply say that Michael killed his sister because she had sex does not explain enough; Michael's motivation must be adequately rooted in a social explanation that elaborates on why her particular sexual activities are specifically punishable by this character. It is also inadequate to say simply that Michael is crazy—his story only becomes of interest to others if it can be shown to have some social value, some wider application of the principles guiding its narrative developments. What, then, could be so compelling about Judith Meyers's behavior that only murder by her brother could set right the chaos she caused?

Michael Meyers finds himself in a position with his sexually active sister of either having to condone her bestial behavior (for having sex outside the family) or risking, if her sexuality is restricted to the family, an incestuous relationship with her. That he actively considers the possibility of an incestuous relation with her is indicated by the voyeurism in the first segment during which Michael vicariously participates in her sexual encounter. Such voyeurism sets up the female as the object of male desire and eventually his violent acting on that desire. Michael moves towards more directly experiencing sex with his sister when he dons the same mask the boyfriend wore during sexual relations. Michael literally puts himself in the boyfriend's place—puts on his false face—and offers his sister a substitute phallus in the form of a knife. The mask that Michael wears during the murder severely limits his vision and the only object that ever enters his visual field is the body of his sister; the protective mask thus also forces him to focus on that which is forbidden to him without it.

The alternative of ignoring his sister's behavior, staying outside the home and resisting the temptation of voyeurism and incest, appears not to have occurred to Michael. Perhaps this is because Michael, as the reigning male representative of the family while his parents are out of the house, is made to appear a fool because of his sister's bestial sexuality. She exchanges her body outside the family in a way that does not provide for continuing social exchanges that Michael and other family members will benefit from (the boyfriend won't promise to call later). The fact that Michael Meyers (he is always spoken of using both his personal and his family name suggesting, perhaps, the importance of his family ties) is wearing a clown's or fool's costume (in red, white, and blue, no less) reiterates his cuckold status.

Michael's elimination of his sister enables him to avoid the dilemma of choosing between condoning bestiality and committing incest. When he does this, Michael is no longer an irrational or possessed boy but a functional adult member of a social unit. An odd visual effect reinforces this transformation of Michael into temporary adult status. When Michael picks up the knife from the kitchen drawer, the camera suddenly rises up much higher and Michael's point of view is now that of a full-sized man.

It is only when he is unmasked by his parents outside their home that Michael sees as a little boy again; or rather, he becomes catatonic and no longer seems to see anything at all. The catatonic state in which Michael Meyers appears to exist for the next fifteen years protects him from the sexual traumas and social decisions of a normal childhood and adolescence. Why then would Michael suddenly decide to escape from the mental institution after fifteen years and return to wreak havoc on his home town? The opportunity for Michael's escape comes when he is being prepared for a trip to a courtroom for an evaluation. Dr. Loomis cryptically explains to the nurse who accompanies him to get Michael that this must be done although it seems a foolhardy act to undertake. The original theatre release of the film leaves this point quite vague and this mystery adds to the tension of the original film.

When the film was being prepared for television and home video release, director John Carpenter reworked the film and made the reasons for Michael's evaluation clearer. In a new scene added to the edited version of *Halloween*, Dr. Loomis is shown six months after the murder of Michael's sister arguing with doctors at a mental institution that Michael should never be released. The doctors explain to Dr. Loomis that Michael will be held for now but that he must be tried as an adult for this murder on his twenty-first birthday and that this reevaluation of Michael's case is the law and beyond their control. If Carpenter found it so important to clarify this point in the doctored versions of *Halloween*, we should take heed that some important information is being foregrounded here.

In the same scene added to the original movie version, Dr. Loomis also explains that Michael Meyers is the most dangerous patient he has ever observed. Loomis tries to convince the other doctors that Michael is covering up, that he is not really catatonic but is just pretending and is hiding an evil other inside. Loomis says, "There's an instinctive force within him. He's waiting." When one of the other doctors asks, "For what?", Loomis admits, "I don't know." When Loomis goes in to see Michael after this discussion he says to the boy who is sitting staring out the window (and who seems to have grown inexplicably because the scene was shot several years after the original), "You fooled them, haven't you Michael? But not me!"

The secret alarm that triggers Michael's escape and return home is his legal status as the sole surviving adult male of his family (the rest of the family has abandoned the house after the murder) who is obligated in a patriarchal society to control the bodily exchanges of his sister. Michael escapes, then, when he has reached his twenty-first birthday and legally become an adult. Being a legal adult imparts to Michael Meyers not only the possibility of legal responsibility for his actions as a young child but also the legal rights of an adult (no matter what his mental capabilities). He can, in a sense, now "legally" do what he had done illegally before—enforce familial sexual protocol on his "sister." Evidence that this film is determined to address the control of familial sexual conduct comes from another scene added to the reworked version of *Halloween* in which Dr. Loomis is shown the word SISTER scrawled in blood on the door of Michael's abandoned room in the mental hospital. Dr. Loomis knows Michael will return to Haddonfield because of this message (the ads for the movie contain the line, "The night *he* came home!"). Michael will, under generic requirements of the horror/slasher film, seek out either his sister or her symbolic equivalent when he goes home.

But his sister is dead and his new intended victim is a seemingly unrelated young woman, Laurie. Pauline Kael, in her review of the film written at its release, says, "for no discernable reason the bogeyman . . . zeroes in on her near the start of the picture but he keeps being sidetracked" (128). It is not true, however, that there is no reason for this choice. In the original movie version, when Michael Meyers first comes back to Haddonfield, the first young women he sees is Laurie, leaving a key under the door mat of his old abandoned family house. He is inside looking at her through the door, and the offer of the key can be taken as simplistic symbolism for a sexual invitation. Then as Laurie walks away from the house towards school, Michael Meyers stands on the sidewalk looking at her back as she sings a song which includes the words, "I wish I had you all alone, just the two of us." Laurie appears to give Michael signals that she is desirous of some contact with him. Michael, who has been described as having "no reaction to external stimuli," reacts obsessively to Laurie's actions by following her and appearing to her throughout the day. As she sits in a lecture during which her English instructor drones on about the nature of fate, Michael appears and disappears as Laurie looks out the window.

While this sexual invitation can be construed as sufficient reason for Michael to stalk Laurie, another more convincing reason is revealed in the sequel to *Halloween, Halloween II.* Here we are given the startling revelation that Laurie is in fact Michael Meyers's real younger sister who was adopted by a local family when their parents were killed. Hill and Carpenter wrote the script for *Halloween II* (although they did not direct or produce it) which is a continuation of the same story later that Halloween evening as Laurie is taken to the hospital and Michael stalks her there. It is impossible to know if this sibling relationship between the two characters was intended in the first *Halloween,* but Carpenter's rewrite of the television and video releases of *Halloween,* created with *Halloween II* in mind, points even more boldly to the possibility of the simultaneous fear of and desire for incestuous relations as a motivating force in Michael Meyers's actions and in the film narrative.

The "familiar," according to the truism (and the often cited biological research of Edward Westermarck [Arens 66]), breeds contempt, that is, a natural avoidance of those who are close. At the age of six and again at the age of twenty-one, Michael Meyers makes a decision to destroy the familiar, his sister, instead of having to make the decision either to remain natural and avoid incest or to become cultural and rational and always be plagued by the possibility of incest. What continues to plague him and what has made him speechless in his fifteen years of institutionalization is the knowledge that he has still not been freed from this choice because another sister's sexuality looms in the future. Michael's return to Haddonfield is not only related to his own coming-of-age but also to Laurie's incipient sexuality and the need to control her body before it is recklessly exchanged. In fact *Halloween IV* offers another female relative (Laurie's orphaned daughter) that he must dispatch before he can have peace. His inability to remain dead is a function not of his supernatural status, but of his newly social body which continues to plague him with thoughts of incest/murder.

Given this information from the sequels, many of the early observations in the film about Laurie and Michael now seem to hint at the possibility of a romantic, and thus incestuous, relation developing between the two of them. When Laurie enters a house to baby-sit, Michael emerges from the shadows while trick-or-treaters chant, "Trick or treat, trick or treat, give me something good to eat." Laurie's friend Annie, pretending that the mysterious man (Michael) that Laurie keeps claiming to see is behind a nearby bush, says to Laurie, "He wants to take you out tonight." Laurie is being teased by her friends because she seems to have no interest in dating while her friends are actively involved in sexual relations. Laurie, like Michael, is a social isolate who avoids human contact. Laurie offers Michael the only opportunity for redemption, the only chance both to retain his relation with the other within and to have touch with an other that is apart from his own body. Laurie's friends must be dispatched because they threaten this relationship. Not only are they giving her improper sexual instructions, but they threaten to stand between her and Michael when they set Laurie up for a date to the upcoming dance with the boy she has a crush on.

But one need not have waited for the revelations of the sequel to understand that the relationship between Michael Meyers and Laurie is a very close one. In many ways Michael and Laurie are the same, not just because they are siblings but also because they are both struggling with their social and sexual identities. Clover suggests that in the slasher film the Killer and the Final Girl share a shifting gender identity that is both masculine and feminine (in *Halloween*, Michael's feminine middle name, Audrey, marks him as sexually confused). Towards the end of the film, Laurie and Michael Meyers change roles as she becomes the "masculine" attacker (stabbing him and symbolically castrating him by blinding his eye) and he becomes the victim. In one very odd moment, after the first time she has stabbed Michael, Laurie sits on the couch, holding a bloody knife above her genitals as if she needs to castrate her now masculinized body. If Michael's body is inhabited by an evil force, so, it seems, is Laurie's and she attends to that horrid force for a brief moment in that scene.

What everyone, including Dr. Loomis, overlooks is the fact that Michael's initial murderous act is against his *sister* and not some other member of his family or some stranger (for example, her boyfriend). Between a brother and a sister in our culture there is always the potential for an incestuous relationship. Sexual relations between primary kin are prohibited to some degree in all societies, although not in the same ways everywhere. According to the anthropologist Walter Arens, most if not all species have prohibitions against "inbreeding" and humans would be different from other animals, or would be "unnatural" if they did *not* have these concepts. What *is* unnatural, what truly marks us as human, is the fight we wage against this "natural" avoidance through our insistence on having incestuous relations despite the "instinctive" and social sanctions against them.

Arens, in an effort to rethink the study of incest, suggests that the prohibitions against incest are not proof of our humanity; rather it is the breaking of that taboo that is proof that we can act beyond our (humanly defined) animal nature. Incest is thus paradoxically both social (involving contact with a forbidden other body but also defining us as human) and antisocial (since that contact, because it is with the "wrong" type of body, threatens the plan for extra-group social exchanges). Michael Meyers was struggling with the simultaneous fear of and desire for incestuous relations with his sister that every Man must struggle with. How a narrative treats the brother-sister (man-woman) relationship demonstrates how successfully this potential antisocial problem has been negotiated. In Michael's case fear got meshed with desire and murderous and sexual impulses crossed over to create a Halloween massacre.

A psychoanalytic explanation might have Michael killing his sister in revulsion and fear after "witnessing" her having sex or after viewing her "castrated" genitals (Dika 97). The film itself suggests, however, especially with the failure of Dr. Loomis's fifteen years of treatment, that there is a better explanation than a psychoanalytic one, one that relates to the fear of those who act antisocial. Loomis says, "I spent eight years trying to reach him and another seven years trying to keep him locked up, because I realized that what was behind that boy's eyes was purely and simply . . . evil." Presumably this judgment is due to the fact that Michael, as a six-year-old, not yet of the "age of reason" and still basically innocent and unacculturated, could not have rationally planned such an evil action. Only if evil was seen as unnaturally inhabiting Michael's innocent body/ mind could Dr. Loomis explain this behavior.

In keeping with this haunted body theme, Sobchack claims that in all horror films terror grows from the fear that we are forever bound to that weak, animalistic part of our bodies which may turn on us any minute and reveal us to be nonhuman (39). This view that we fear the animal aspects of our bodies does not account for the social value of such terror. I want to suggest that the real horror results not from the fear that we could revert to being animals; this is something we already know. Instead, the horrible thing to acknowledge is that we are, indeed, stuck with being human with socially defined bodies torn between the conflicting desires, prohibitions, and temptations of our community. And, as the Colony in *The Howling* suggests, it is better and easier to admit that one is not really and truly *just* a human being, because it is the constant human struggle to try to maintain a facade of social order that causes such perpetual human angst.

The most horrible thing about being revealed to be both human and nonhuman is that we can also be exposed as potential antisocial threats to the system of social exchanges. We may be left out of the cycle of gift exchanges if this happens. Incest is antisocial because it takes out of circulation bodies that are supposed to be involved in socially productive exchanges. It puts these bodies into relations that feed only on themselves, making them, in a sense, cannibalistic. When a body is inhabited by an evil other, it is creating its own isolated social world, creating its own rules and keeping social relations in a discrete but perverse unit. Under this condition we become other to ourselves, a most horrifying state of madness in which we experience not just disgust at our psychological or physical transformation, but horror at our falling away from the human world where social exchange—the gift of touching others—determines our humanity. Michael's isolation as a catatonic in a mental institution effectively separated him from all social contact—being catatonic is a state of failed reciprocity. When a doctor remarks after Michael's escape that he could not possibly have driven a car away because there was no way he could have learned to do this, he acknowledges Michael's isolation from real world contacts. Yet Michael does drive the car away, suggesting that he has not been as isolated as he pretended, that an other has been keeping him company.

On Halloween night, a time of ritual reversal, people put on masks and costumes and take on roles they normally would not play. The world is turned upside-down as men become unsavory characters or women, women become men or children, children become fantastic creatures or animals, the dead walk, and many of the living pretend to be deceased. Costumed participants and trick-or-treaters are showing all those things that can destroy society, those things that all the rest of the year must be kept under control if society is to continue. In a liminal period all the rules that usually govern ordered social life are suspended. In the practice of costuming and masking, people are trying out these other, forbidden roles for a night. They can taste the forbidden sins without the serious long-range sanctions as long as they do it during institutionalized times of upheaval. Michael is his own Halloween costume, his own evil other. Michael switches on Halloween from his catatonic asocial self to his murderous, lusting antisocial self. He chooses Halloween night for his murders because he, too, is addressing those things (incest and bestial sex) that can make society fall apart. But because he has no stake in reintegrating society after this night of reversal, his acts are irreversible (murderous) ones.

As a time of masking, Halloween suggests the falsity of social roles in the same way Mauss suggested that all social exchanges involve deception. Everyone in Haddonfield wears some kind of mask. The younger Michael wears a mask when he is murdering, and the older murdering Michael not only wears a mask himself but also puts one on some of his victims, suggesting they are either being punished for being false, or that they stand in for a larger category of person and their individual features are not important. Annie and Linda, Laurie's two friends, use their babysitting as a mask for sexual encounters with their boyfriends. Both Annie and Laurie have to pretend to be innocent of smoking marijuana when they pull over to talk to Annie's father, a policeman—but only Laurie worries about how they appeared to him afterwards. Dr. Loomis at one point hides behind some bushes to scare children away from the old Meyers house just as little Tommy, the boy Laurie baby-sits for, hides behind some curtains in order to scare Lindsey, the little girl Annie is supposed to be babysitting—each pretends to be a frightening creature in order to control his situation.

To be constantly masked may seem socially irresponsible because it appears that appropriate social roles are not being properly conducted and have been replaced by deceptions. But the fact that Laurie, who is supposed to be the most honest and innocent character in the film, wears the most complex set of masks suggests that this is one of the requirements for living in communion with others. Laurie, not yet being truly social/sexual, is confused about the need for these deceptions. She pretends she doesn't think about boys although she secretly has a crush on one in her school; she says boys don't like her because she is too smart and she has not yet learned to hide this side of her; she attempts to smoke marijuana with her friend Annie but only succeeds in coughing and grimacing; she carries a pumpkin in a way that makes her look like she is pregnant; and she wears an apron while she baby-sits which allows her to pretend to be a mother. Laurie is continuously trying out different social roles, a practice that makes her dangerous and in need of Michael's guiding hand.

Halloween, according to a *Newsweek* review at the time of its release, "has no socially redeeming value whatsoever" (Ansen 116). This review is quite wrong, however, because as we have seen the cinema is a gift whose enormously important social value rests in its role of providing not only guiding mythological stories but a social obligation to respond positively to those tales. The mutual disciplining of the film by the audience and of the audience by the film (especially through the use of genre stereotypes) makes films' productive social exchanges worth considering seriously no matter what their silly plot or their technical merit. The power of horror films may rest precisely in their function of inviting into the social world of exchanges those persons, particularly adolescents, who need to learn about societal requirements on bestiality and incest. But rather than just instruct them, these films' texts obligate them to behave in a particular way. For the adolescent audience, and indeed for the aging adult aficionado, examples of and reassurances about bodily changes can be found in horror films which enact all the possible bodily boundary transgressions, transformations, and their consequences. What the films ask in return, however, is even more horrifying.

As Mauss explains in discussing the dynamics of gift exchange, "We must always return more than we receive; the return is always bigger and more costly" (63). Georges Bataille, in his impressive extension of Mauss's ideas, claims that extensive and elaborate gift-giving ceremonies like the northwest coast tribal "potlatch" have as their goal, "humiliating, defying, and *obligating* a rival" (121, emphasis in original). In the potlatch, the donor either gives away or destroys the vast quantities of material goods that are a sign of his accumulated wealth. This is often done as part of a rite-of-passage or ritual ceremony. He elaborates that, "the exchange value of the gift results from the fact that the donee [the receiver of the gift], in order to efface the humiliation and respond to the challenge, must satisfy the obligation (incurred by him at the time of acceptance) to respond later with a more valuable gift, in other words, *to return with interest*" (Bataille 121, emphasis mine). This interest can be considered not only in the usury sense but also in the social sense—the potlatch/gift exchange creates an interest in keeping the cycle of social interactions continuing. The cinema can be considered very much like a potlatch—we are offered extravagant gifts that are presented and then destroyed right before our eyes when the lights go up. We leave the theatre, as Bataille would claim, stunned, humiliated, and obligated to create our own matching offer.

Our obligation in return for these horrible stories of broken taboos in the horror film is the total surveillance of our bodily activities so that not only will incest and bestiality not occur, but so that no activities that might incite or inspire others to these acts will be allowed to take place. In a continuing gift spiral, we obligate ourselves to increasing self-control through movie-viewing, and we therefore obligate the movie industry to give us more and more tales of "illicit sexuality" that will thankfully remind us of our weaknesses, our potential sins, and the thin line between real life and horror.

I have proposed that the overwhelming social sanctions about incest and bestiality that Michael Meyers has not yet learned how to deal with are the motivating factors that trigger his original actions and his repeat performances in the *Halloween* films. An examination of the contradictory and compelling fear of and desire for "incestuous" and "bestial" relations that motivate not only this film but much of social life has shown how the need to maintain proper social exchanges is more important than personal psychological states for explaining the way bodies are used in horror films and in the societies compelled to make and view them. Furthermore, the anticipated and necessary failure of film characters and viewers to learn this lesson results in endless sequels and spin-offs that perpetuate the exchange of moral lessons and proper behavior needed to govern unruly bodies in human societies.

WORKS CITED

Ansen, David. "Trick or Treat." *Newsweek* 92 (4 December 1978): 116.

Arens, W. *The Original Sin: Incest and its Meaning.* New York: Oxford University Press, 1986.

Bataille, Georges. *Visions of Excess: Selected Readings, 1927–1939.* Ed. Allan Stoekl; Trans. Allan Stoekl, Carl R. Lovitt, and Donald M. Leslie. Minneapolis: University of Minnesota Press, 1985.

Clover, Carol. "Her Body, Himself: Gender in the Slasher Film." *Misogyny, Misandry, Misanthropy.* Eds. R. Howard Bloch and Frances Ferguson. Berkeley: University of California Press, 1989. 187–228.

Dika, Vera. "The Stalker Film, 1978–1981." *American Horrors.* Ed. Gregory Waller. Chicago: University of Illinois Press, 1987. 86–101.

Kael, Pauline. "Halloween." *New Yorker* 55 (19 February 1988): 128.

Mauss, Marcel. *The Gift: Forms and Functions of Exchange in Archaic Societies.* Trans. Ian Cunnison. New York: W. W. Norton, 1967.

Prince, Stephen. "Dread, Taboo and The Thing: Toward a Social Theory of the Horror Film." *Wide Angle* 10.3 (1988): 19–29.

Shell, Marc. "The Family Pet." *Representations* 15 (Summer 1986): 121–153.

Sobchack, Vivian. *Screening Space: The American Science Fiction Film.* 2d ed. New York: Ungar, 1988.

4

Olfactory Constitution of the Postmodern Body: Nature Challenged, Nature Adorned

INTRODUCTION

> People could close their eyes to greatness, to horrors, to beauty, and their ears to melodies or deceiving words. But they could not escape scent. For scent was the brother of breath. Together with breath it entered human beings, who could not defend themselves against it, not if they wanted to live. And scent entered into their very core, went directly to their hearts, and decided for good and all between affection and contempt, disgust and lust, love and hate. He who ruled scent ruled the hearts of men. (Suskind 189)

> It's theirs. It's personal. It's not their husband's, not the kids', not the neighbor's, not their co-workers'. It's nobody's, it's

theirs . . . it's really their very own, and no one else can have it. Their father can't take it away . . . it's like no one else on the whole planet wears it . . . "It's mine. I own this. This is me." (Interview, body-artist/perfume designer)

In keeping with the popular practice of appending test-strips to perfume magazine ads, I use two narrative-strips to open this essay about stories that fragrance narratives inscribe upon female bodies.[1] Each strip speaks to what are actually linked themes in many fragrance narratives, one evoking a naturalized body subjugated to biology, the other evoking the self-authoring subject whose body is "made" hers. In reading fragrance narratives here, I take the body to be a site where notions of nature and culture are encoded and contested. But I am also concerned with questions of subjection, subjectivity, and agency, and the extent to which the body so conceived affords a "new" postmodern politics of resistance.

These opening narrative-strips, which draw upon notions of a body-identity both fixed in nature *and* open to inscription in culture, thus serve to introduce one argument of this essay: that contemporary fragrance narratives inscribe bodies as "naturally" female in a singularly "cultural" way. As importantly, they foreground the discursive frames within which such bodies are constituted. They thus stake out the ground where we can look for resistance to this constitution of bodies, and where the political consequences of these engagements themselves can be evaluated.

In the case of the "naturalized" body, the first narrative-strip promotes scent as being "of the body," as bodies give off scents and scents are detected and responded to without conscious intervention. Alluding to core substance and uncontrolled—"uncultured"— urges and drives, it draws upon an evolutionary biological discourse which, we will see, constitutes female and male bodies as absolutely different, each following distinct "natural" heterosexual drives and urges. In this, it is like other narratives that I consider below: psychological and scientific narratives about scent and sexuality, theoretical narratives about romance, female eroticism, and desire.

The second narrative introduces a vexing twist to the scent-body link, arguing for the possibility of negotiating the positions that can be taken up vis-à-vis other subjects in a uniquely authored body.[2] In this case, what is being called upon is a postmodern sensibility which refuses the fixity of a unitary subject in favor of a self-authored, multiple subjectivity; it acknowledges, if not celebrates, the constructed nature of the world.

While hardly constituting a discourse on the order of the discourse of evolutionary biology, such postmodern sensibility is nonetheless lived, and increasingly so, in postmodern culture. In an abstract sense postmodern culture is one in which "images are consumed as simulations of social identities" (Angus and Jhally 5). In more concrete terms it is, to cite part of a richly panoptic characterization,

> the shopping mall, mirror glass facades, William Burroughs, Donald Barthelme, Monty Python, Don Delillo, *Star Wars*, Spaulding Gray, David Byrne, Twyla Tharp . . . the Hyatt Regency, the Centre Pompidou, *The White Hotel, Less Than Zero*, Foucault and Derrida; it is bricolage fashion, and remote-control-equipped viewers 'zapping' around the television dial. (Gitlin 347–348)

The last image of "viewers 'zapping' around the television dial" is particularly useful as we seek to untangle the link between representations of fragrance—what we understand it to be, how we understand it to work, how we use it—and the constitution of a female body that is imbued simultaneously with natural signals and drives and with the agency of the postmodern subject. For it reminds us that consumers of fragrance and fragrance narratives operate increasingly in a world characterized by instantaneous media interpretations and representation of events and things.[3] In this context consumers may come to understand images as constructions based upon multiple media interpretations while simultaneously responding to them as representative of some "original that supposedly precedes simulation" (Angus 344).

To discern this interplay in the constitution of a female body, through images of fragrance and fragrance use, I draw below upon personal and published accounts of women who use perfume as well as upon texts generated by producers and marketers of perfumes. These texts include print and television advertisements for perfume, and papers delivered at a 1986 international conference of perfumers, manufacturers, and perfume researchers. Unique among the accounts from within the perfume industry is the story told to me by a self-styled body-artist about her creating, packaging, and marketing of a leading "celebrity perfume."[4] Collected over several interview sessions, it offers the richly detailed and intimate perspective of an individual directly involved in "thinking up" and producing images that are called upon in inscribing a postmodern female body.

In taking up this reading of fragrance narratives, I have, finally, been concerned to explore scholarship which offers the contestatory power of parody (Butler), doubling (Smith), travesty (Schor), and exhibitionism (Silverman), for these hold out the promise of a political practice that would shatter dichotomies, explode the cultural underpinnings of naturalized bodies. I have been drawn in particular to Haraway's image of the iconoclastic, resisting, monstrous cyborg body that:

> is not innocent; . . . was not born in a garden; . . . does not seek unitary identity and so generate antagonistic dualisms without end (or until the world ends); . . . takes irony for granted. ("Manifesto" 99)

Among the perfume narratives that I read here, there is one in which such a monstrous body figures centrally. This is Suskind's postmodern novel,[5] which is aptly titled *Perfume*.

In the opening scene of *Perfume*, we are witness to the horrific birth of the central character, Grenouille, from the fetid body of a mother who pushes him into a pile of rotting garbage to die. While the mother figure is depicted as someone to be feared and despised, the central character, too, is presented to us as abject: Giving off no natural smell of his own, he repels his wet nurse; born male, but "in lack," Grenouille's subjectivity within signification is up for grabs. Grenouille's experiences as his story unfolds literally embody these impossible tensions. At once striving to constitute himself as a natural human subject by inventing a "normal" scent for his body, Grenouille at the same time seeks the disruptive power of the multiple self, in a scent distilled from the many different bodies of murdered maidens.

The erosion throughout *Perfume* of key distinctions—between subject and object, between humanity and monstrosity, between male and female—calls into question notions of "nature," "culture," and the gendered body. I have thus found much in *Perfume's* narrative useful in exploring the potential for contestation in women's fragrance use and in their consumption of images promulgated through fragrance narratives.

Of course, fragrance and narratives about fragrance and the female body are not alone in being open for consumption. In aligning this essay's use of opening quotes with a popular perfume marketing device, I underscore my concern to expand the scope of this essay to include a reading of the body as it figures in the production and consumption of postmodern theory as well as postmodern culture (see also Mascia-Lees, Sharpe and Cohen). In this sense, too, I have found *Perfume* to be useful. For while engaging the reader with the promise, in subversive postmodern fashion, of exploding categories, the novel's narrative at times also reproduces the very categories it would explode. The first opening narrative-strip evoking a natural body, for example, is taken from *Perfume.*

As the "ground" for fragrance narratives about both biology and agency, then, the body is another element to be read here, and in particular as it is constituted and deployed in postmodern scholarship and politics. Thus, in the concluding section of this essay I read perfume narratives with an eye to deciphering "the postmodern body" that, understood as culturally constructed, has figured so centrally in efforts to contest sex and gender as essential categories in the constitution of human subjects.

FRAGRANCE AND THE CONSTITUTION OF A FEMALE BODY IN NATURE

In configuring the scent/body relationship in postmodern consumer culture, narratives about scent and fragrance deploy multiple and complex messages and codes about a female body. Organizing these messages and codes is a particular and historically specific discourse about nature that arranges male and female bodies in a system of absolute difference. In this discourse the male body—its physiology, its sexuality, its desires—is established as the human subject, active, "in culture." The female body from which this human body is differentiated is the body "in nature," subject to species reproductive needs, the passive object of male sexual desire.

We can see this discourse in operation in the narrative of the body-artist/perfumer. Directly after invoking an image of Cleopatra "oiling her skin" to "lure her lovers," this narrative asserts:[6]

it's nothing new, to lure their men. I mean, that's how animals do it. . . . They secrete a distinct odor and that's how dogs get into heat. They follow these women into heat, these poor dogs. If you have a female dog there's no more to tell. All of a sudden you're looking out your door and there's fourteen guy dogs standing at your door going "how did I know she's like having her period?" How do these dogs know this? Did she whistle, did she call them on the phone? These poor dogs.

Although the body artist's narrative wants to distinguish fragrance as something "put on" from odor (scent) as something "put out," it conflates the cultural agent who oils her skin with the natural body which lures and reproduces. This conflation is exposed in the ready substitution of "women" for "female dogs," "guy dog" for "male dog," and "period" for "heat." The unproblematic substitution of these terms reveals an underlying assumption that scent operates in women's bodies naturally, as it does in animals' bodies.

As Thomas Laqueur has recently demonstrated, such an assumption derives from the specific understanding of male and female bodies promulgated by Western biological science beginning in the late eighteenth century. Operating to reestablish hierarchy in the face of the Enlightenment push toward equality, the new "biology of incommensurability" (Laqueur 3) conceived male reproductive and sexual physiology as being responsible for and reflective of male political and intellectual superiority. In contrast to an antecedent understanding of the female as "a *lesser* human, less hot, less spirited; but . . . not a *different* human" (Haraway, "Investment Strategies" 148), the new biology accorded women no function similar to that of men. Rather, denied an interest in or capacity for sexual fulfillment, women and women's bodies were understood to operate strictly in service to species reproductive needs. To come full circle back to the body-artist's narrative about fragrance, this new understanding of female bodies derived special credence from scientific studies that claimed female sexual and reproductive physiology to be like that of animals, particularly female dogs; as Laqueur notes:

> The American physician Augustus Gardner drew out the implications . . . "The bitch in heat has the genitals tumified and reddened, and a bloody discharge. The human female has nearly the same." "The menstrual period in women," announces the *Lancet* in 1834, "bears a strict physiological resemblance" to the heat of "brutes." (27)

In contemporary discourse, such an understanding of differential bodily function is reproduced in sociobiological narratives which constitute female sexuality as naturally coy and passive, male sexuality as aggressive and active. As Haraway submits, sociobiology offers "no story about happily complementary heterosexuality" ("Investment Strategies" 154). Rather, the story told about males and females in sociobiology is one of opposition and competition. According to this story, females' needs to safeguard what are theoretically limited, energy depleting, but high-quality sex-cells makes it "more profitable for females to be coy, to hold back . . ."; males' theoretical ability to "inseminate thousands of women in [a] lifetime" means that "it pays males to be aggressive, hasty, fickle and undiscriminating" (Wilson 124, 125).

Not surprisingly, stories that are told about the relation between male bodies, female bodies, scent and fragrance in contemporary scientific narratives draw heavily upon the sociobiological account of human cultural and sexual evolution.[7] For example, zoologist D. M. Stoddard attributes the "sex appeal" of perfume to the survival needs of a pair-bonded and differentially able reproductive couple (the female seductive, the male aggressive). Positing as selectively advantageous any feature that would "reduce[] the chance that a female might cuckold her bonded mate and . . . curb[] a male's likelihood of philandering" (11), Stoddard argues that the natural mammalian reliance upon steroid-based sex attractants (pheromones) was repressed in early humans. Moreover, while such repression was sufficient to ensure "the retention of the nuclear family within the framework of a gregarious life-style" (11), it was not so complete as to divest pheromones of *all* their power. Indeed, concludes Stoddard, the effect of the contemporary use of "sex attractants" in perfumes is:

> to jog the ancient memory traces in the brain . . . and in a sublime and vicarious manner reveal precisely what the perfume helps to conceal . . . the memory traces of long ago . . . are not so tightly entombed that the subconscious mind cannot be stirred to excitement and provide a mood of attentive satisfaction. (15)

FRAGRANCE, PHEROMONES, AND THE NATURALIZATION OF SEXUAL DIFFERENCE IN "POPULAR" NARRATIVES

In attributing to pheromones emitted by the female body a "male-exciting" function, the above scientific account is consistent with a biological science which posits two sexes, essentially different yet uncontrollably mutually attractive, in service to species reproductive needs. A recent *Cosmopolitan* article reporting on the latest "scientific research" on sexual attraction—and entitled "The Scent of Love"—operates similarly as a thinly disguised rationale for male sexual aggression and female sexual passivity. Describing pheromones as "special aromatic chemicals secreted by one party to affect the sexual physiology of another, to kick off sexual interchange and intercourse," this account depicts female pheromones as floating "diffuse[ly]" through a room, male pheromones as "requir[ing] intimate contact" to be effective. Further, this narrative constitutes male bodies with the agency of the Enlightenment subject, attributing to male pheromones the role not just of driving, but of improving, female reproductive function:

> women who have sex with men at least once a week are more likely to have regular menstrual cycles and a more benign menopause than celibate women and those who have infrequent intercourse. The crucial element, aside from sexual intercourse itself, appears to be exposure to aromatic chemicals secreted in a man's normal body odors. (Orenstein 204)

We see this discourse about male and female sexual difference reinforced in several contemporary perfume magazine ads. For example, *Oscar de la Renta*'s invitation to "experience the power of femininity," calls upon the discourse of a natural sexual attraction of males to females. A close-up photograph which foregrounds the naked shoulders and neck and reddened lips of a simultaneously demure (passive) and sexy (alluring) woman reminds the user of this perfume of her natural role as attractor. In other ads, it is the practices that the story of human evolution has us associate *with* heterosexual attraction that are foregrounded, as in the *Coty* ad that explains a woman's being surprised with a kiss on the neck by a man thus: "It must be the wild musk," or the *Obsession* ad showing a naked man carrying an unresisting woman away on his shoulder.

The magazine ad for the designer perfume *Montana* resists employing bodies in its imaging of sexual difference; instead a simple picture of its double-helix shaped bottle alludes to the genetic basis of sex difference. Fashion magazine copy reiterates the genetic symbolism, heralding this perfume as "part of a return to real perfume, 'with a message' " (*Vogue*). Finally, in what I think can be read as an allusion to stories about the primate in the human female, is an ad for *Opium*. This ad pictures a woman lying upside down, her naked draped buttocks evocative of a kind of primate display—a sort of "Simian Orientalism" (Haraway, *Primate Visions*) inscribed back upon a successor "other."

The stories being offered in these popular narratives do not contest accepted notions of the female body. They are seamless as they impart little sense of discontinuity between a natural sexuality and the sexuality that will be constituted through the inscription of the narrative, by the application of the perfume, upon the body. Like contemporary scientific accounts equating the action of perfume with the action of pheromones or "sex attractants," they maintain and regulate female and male sexual difference as it is constituted "in nature." Indeed, so isomorphic is the relation that they posit between fragrance and female body function in the service of male sexual urges and species reproductive needs that they effectively disallow the fact of other sexual visions and practices in which, as Butler reminds us, "gender does not necessarily follow from sex, and desire, or sexuality generally, does not seem to follow from gender" (136).

And yet the very existence of what Butler has termed "gender discontinuities" (134) suggests that sexual practice is constituted through negotiations taking place among competing discourses. With specific regard to possibilities that are opened up by scent, it seems that the constitution of a reproductive female body driven by male-centered sexual urges is made more problematical when fragrance use or fragrance narratives engage notions of female eroticism, pleasure, and desire. This is because these notions themselves draw upon discourses that posit female sexuality, embodied as libidinal eroticism, as dangerous and disruptive.

FRAGRANCE AND LIBIDINAL FEMALE EROTICISM AND SEXUAL PLEASURE

The disruptive potential of fragrance narratives is in part a factor of an association of scent with the erotic, the libidinal, the pre-symbolic. "Naturalized" by Freudian theory, the libidinal is the realm of the phantasmic. As we see in the following Freudian reading of the psychological impact of fragrances, the psyche is conceived as the realm where desire and object of desire are fused:

> The indrawn breath that accompanies the perception of odors, themselves experienced "inside" the body . . . favor[s] the phantasy of the incorporation of the object by means of its odor, and thus to enhance the magically aphrodisiac properties

of perfume. And the habit of many people who shut their eyes when deeply smelling a "good" odor is . . . directed toward the furtherance of an incorporation phantasy by the deliberate exclusion of the visual perception of the object. (Pratt 83)[8]

Closely associated in Freudian theory with this realm, women are seen as more susceptible to "a relinquishing of control by the conscious mind in obedience to the demands of the unconscious sexual appetite . . . which increases in strength the lower we go in the animal world" (Daly and White 80). Thus echoing sentiments of late eighteenth-century biologists, Freudian psychologists writing on smell see women as being more "impressionable during the menstrual cycle," and "especially liable to erotic hallucinations during anaesthesia" (Daly and White 81).[9]

These associations—of fragrance with the libidinal, with female eroticism, with a relinquishing of control by the conscious mind—are drawn upon extensively in Suskind's novel, *Perfume.* Here, the scent that most attracts the main character, Grenouille, is the scent of the maturing maiden. Convinced that fragrance has the power "to rule the hearts of men" (189), Grenouille heinously murders maidens so to capture and distill their special scent. Descriptions of this scent as "a bud still almost closed tight" and "terrifyingly celestial," (207)[10] establish it as linking a nascent, yet uncultured, female sexuality with an unearthly and frightening power. The precise nature of this power is revealed in a scene of total moral and social disruption that ensues when Grenouille, about to be hanged for murder, appears wearing the distilled scent of multiple murdered maidens:

> It was as if the man had ten thousand invisible hands and had laid a hand on the genitals of the ten thousand people surrounding him and fondled them in just the way that each of them, whether man or woman, desired in his or her most secret fantasies . . . they fell down anywhere with a groan and copulated in the most impossible positions and combinations: grandfather with virgin, odd-jobber with lawyer's spouse, apprentice with nun, Jesuit with Freemason's wife—all topsy-turvy, just as opportunity presented . . . It was infernal. (291)

These associations of a libidinal female eroticism with a challenge to the control exerted by concrete structures of society and culture are, finally, analogous to those that were brought into play in the sexual politics of some strands of American feminism in the 1970s. Here, struggles for equity and autonomy were waged at least partly in terms of women's claims to their orgasmic pleasure.[11] In one sense, this move can be read as seeking to decenter male sexual power and confidence by playing upon prevailing cultural notions of women's desires, with female orgasm standing for the unfettered, unsubjugated female body.[12] In another sense, as Haraway has recently argued, it can be read as a move by white middle class women to disrupt the narrative of the singularly male prerogative of "self-containment and self-transcendence simultaneously" through orgasm. In this scenario, "orgasm on one's own terms signified property in the self as no other bodily sign could" ("Investment Strategies" 150).

In speaking about the meaning of their perfume to them, women draw upon similar notions of the power of female sexual eroticism. While seldom cast in explicit terms of a politics of equality, these women's testimonials nevertheless refuse to replicate in absolute terms the narrative of a male-centered female sexuality. Instead, and perhaps even mobilizing images of female empowerment which are the legacy of the above feminist politics, they evoke a connection between female sexual pleasure and a sense of personal empowerment. In reading these testimonials as well as other types of fragrance narratives that seem to be arguments about female sexual pleasure and power, we can begin to explore the extent to which they actually serve to disrupt the female body inscribed within the discursive frame of a biology of incommensurability.

AGENCY, COMPLICITY, AND THE REGULATION OF THE FEMALE BODY

Narratives of perfume wearers and manufacturers alike speak to a close relationship between women perfume users and their perfumes. In the body artist's narrative, perfume is configured as "the closest thing to your skin . . . there's nothing closer." This is echoed by a perfume wearer who declares of her perfume, "I feel naked without it" (Gordon 55). The connection between a highly personal association with fragrance and a sense of a sensual self is intimated in popular accounts as well, as in a recent article in *Working Woman* magazine:

Our fragrance selections are highly personal. They're often determined by the message a singular scent sends to our psy-
ches—the inner sense of confidence and "specialness" it inspires. (Resnick 120)

Again, we see this connection being made by a perfume user herself:
"I put on perfume before I go to bed no matter whether I am with a
partner or not. I find it very sensual" (personal communication).
Noting, finally, that "scent can . . . be a strong confidence builder,"
(Mark 19)[13] psychological studies of perfume wearers make a direct
link between fragrance use, a sense of personal empowerment, and
sexuality:

We find perfume being used to give the wearer confidence that
the desired sexual gratification will be attained . . . this excite-
ment appears to be at least as strong an anticipatory factor as
the actual sensual experience. The ritual of "getting ready,"
whether for an evening out or in bed, is important. (Byrne-
Quinn 206)

These narratives are alluding to a consciousness on the part of
the consumer and manufacturer alike of a highly personalized sensuality associated with fragrance. In this sense, they seem to evoke a
female desire and pleasure apart from a discourse about male-centered sexuality and reproductive sex. And in this sense as well, they
would seem to leave an opening for the constitution by women of a
subject position in their own self-contained desire.

But such constitution does not, in fact, appear to be entirely
possible. For in no sense is the above sensuality or confidence linked
with the sensual experience of the perfume wearer's satisfaction of
her *own* desire, as would be the case, for example, of clitoral
orgasm.[14] Rather, as the narrative of the above researcher later
makes clear, the perfume wearer is never considered to be directly
engaged in the fulfillment of her own desire:

The direct response is concerned with the effect on the wearer.
But the effect being striven for in the perfume "message" is the
effect of the wearer of the perfume on someone else. (Byrne-
Quinn 206)

Narratives of perfume wearers express a contrast between a deeply felt sense of personal pleasure in their perfume and a sense that this pleasure is open to different interpretations than the one they might wish to give it. They thereby give voice to this underlying contradiction between female eroticism and pleasure as an expression of self-containment and female eroticism and pleasure as they stand to be appropriated in the sexual interest of another (male) subject.

For example, a recent women's magazine article reports that one woman who admits deriving pleasure from what she characterizes as particularly sensual perfume would not wear it to an important meeting because, "[f]ragrance can be a distraction, like wearing a blouse unbuttoned. It automatically says you're a woman." Another woman is reported to have stopped wearing perfume because, "I work in an academic setting dealing with human sexuality, so I don't wear fragrance at work. I don't think I'd be taken seriously if I did" (Gordon 19). And in a narrative that refers as well to a subliminal male sexuality, we see a recent "User's Guide to Fragrance at the Office" cautioning working women to "think of properly applied perfume as background, not center stage. While you may not be continuously aware of it, others will be." Moreover, this article submits, while perfume is "no substitute for substance," it can be deployed effectively in the building of a professional image. The key to this is "knowing how to use it" (Resnick 122).

This "User's Guide" narrative is particularly revealing of the contradictory subjectivity that is being constituted in fragrance use through reference to the notions of female eroticism and pleasure that are in play. On one hand, referring explicitly to a female eroticism, it acknowledges and foregrounds, as do the preceding narratives, the possibility of a female subjectivity constituted in sexual pleasure. Indeed, in this latter narrative the female subject so constituted is even attributed with agency: she can "know how to use it." In this sense, this and related narratives would seem to offer women a position from which to contest the natural heterosexual, male-urge-driven reproductive female body.

We see, for example, claims being made for control over what is sexually pleasing, and when, in some women's self-reflective contestations of what they perceive to be traditional perfume practice. A president of an executive search firm thinks that fragrance is "so special, so feminine and so sexy that I purposely don't wear it to work. I feel wearing it makes a personal statement that is so strong I reserve it for private moments and social occasions." Similarly, a merchandise director of an exclusive retail store reports never leaving the house "without putting on fragrance. I love wearing it to work because it *is* a reminder of the rest of my life" (Gordon 19).[15] Keeping pace with these sentiments, a woman in a recent television ad for *Luna Mystique* perfume claims that "when it touches me it's as if the power of the moon is mine and I can have all that I desire."

On the other hand, in cautioning the perfume wearer that, even if *she* isn't, *others* will be "continuously aware" of her fragrance, the "User's Guide" narrative evokes a natural female body, subject without consciousness to the male sexual desire that its scent may unleash. Thus, while acknowledging female eroticism and even encouraging women perfume users to enhance and deploy it strategically, this narrative simultaneously confines these strategies to a sexual realm within which women have no real control. Within this realm of heterosexual attraction, it is, in fact, women's naturalized bodies that talk.

Quite early in the contemporary feminist critique of consumer culture, Firestone argued for an understanding of romance as constructed "to preserve direct sex pleasure for the male, reinforcing female dependence," fulfilling women sexually "only by vicarious identification with the man who enjoys them" (167). At the time, this argument served as one of the theoretical impetuses to practices which deployed the clitoral orgasm in a struggle for women's sexual autonomy and subjecthood. But today, references to women's sexual pleasure are incorporated into ads selling perfume and are highlighted in testimonials of women who buy and use perfume. Indeed, in this postmodern era in which claims to female erotic potential are part of popular cultural discourse, what is being constituted as romantic is simultaneously a sexuality over which women take responsibility *and* a sexuality capable of calling forth, without conscious intervention, uncontrollable male urges.[16]

This envisioning of the romantic is played out in the conclusion of the *Luna Mystique* perfume television ad which evokes both a power beyond a woman's control—"if the moon can move an entire ocean"—and individual agency—"imagine what it can do for you." Similarly drawing upon this image of romance is the magazine ad for *Impulse* body spray. The ad focuses our attention upon a woman who is sitting/leaning comfortably upon a convertible with perfume spray at the ready. She appears to be entirely in control of a young man who, back to us, is about to present her (romantically) with a bouquet of flowers. This power of sexual self-inscription is acknowledged in the ad's written narrative: "A different scent for every day of the week. Fresh. Free-spirited. Spray *Impulse* body spray all over *to create a sensation of your own"* (italics added). This power is held in check, however, and romance as a sexuality over which women have no control is asserted through the bold-print headlines for this ad: IMPULSE. BODY SPRAYS. SEVEN SENSATIONAL WAYS TO MAKE HIM ACT ON IMPULSE.

In this same vein, but even more provocative, is the magazine ad for *KL,* a perfume "designed" by Karl Lagerfeld for Marshall Field's. This ad foregrounds a woman in an evening dress, her back against a closed door. She is looking at a naked man approaching her, but her face is empty of any singular emotional content. The ambiguity that we encounter in attempting to interpret this woman's look is furthered by the use of black and white photography and stylistically confusing accessories—a fan evocative of fashions of the 1930s, a "simple black dress" evocative of the sensibility of the 1950s. In a manner characteristic of "mode retro" films, the ad is thus emptied of substantive content, and any salient distinctions (as between, for example, nature and culture) are confused (Jameson 1983). A single line of text appearing directly above and to the right of the woman warns, EXPECT THE UNEXPECTED. Again, like the confusion in the semiotics of this ad, its written narrative, which takes the form of a warning, simultaneously invites and accuses: it is the woman, having responded to the cue "expect" by putting on the perfume, who has provoked "the unexpected."

Because they draw upon stories that we tell ourselves about natural and "uncontrollable" sexual instincts, either of the above ads could have been readily included in the preceding section of this essay. As narratives of sexual difference, they call upon and reinforce notions of male sexuality as instinctively and aggressively responsive to the subliminal seductive sexual cues of the female. But, as we have seen, even as they subject women to the "uncontrolled" forces of male sexual nature called forth by the natural female body, these narratives portray women as self-constituting subjects, capable of resisting such an essentialist inscription by authoring their own body-texts. These ads thus reveal another order of complexity in the natural female body as it is put into play in the postmodern context; this is the complexity introduced by their reference to the natural libidinal female body.

For these narratives establish women's capacity to resist through a link between the libidinal female body and the disruptive power with which it is associated. This was also the argument of the *Perfume* excerpt and, as the body politics of 1970s suggest, it is a link that has some currency in popular discourse and practice. But linking the agency of the thinking, strategically decisive subject to the natural powers of a libidinal body activates as well the body constituted in nature in service to species reproductive needs. And as *it* is constituted, this natural body has the task of unleashing, without conscious intervention, the uncontrollable sexual drives of the embodied male. As soon as it is evoked, then, the libidinal body is also opened to subversion to the interests of the reproductive body, even as it grounds claims for self-conscious action on the part of women.[17]

It is this subverted libidinal body that these perfume narratives ultimately constitute: a body engaged in complicity with nature *against* any claims to the possibility of its inscription outside of nature. Thus, as in the case of rape—and especially so in cases of date rape—the very acknowledgement of a female desire or eroticism that is open to manipulation and capable of disrupting sexual continuities is deployed in support of arguments that hold women responsible for the violence that is done to them and especially violence done upon their bodies.

FRAGRANCE AND THE RESISTING MULTIPLE SUBJECT

In Volume I of *The History of Sexuality,* Foucault submits that the sexed body is the regulated body, subjected as with physical restraints by a historically specific discourse about sex and nature. Following from the understanding of the sexed body as the body regulated, Wittig argues for the destruction of sex as a category altogether; here, the subject who would begin from "I" (not I as standing for the universal/man, or I as standing for woman), would become "*the* subject through the exercise of language . . . an absolute subject" (66). While not explicitly arguing against the sexed body, Cixous too holds out the revolutionary potential of the body that is a celebration of erotic multiplicity. Similarly, Kristeva locates the power to disrupt the body as it is encoded in the symbolic (the heterosexed body) in the pure *jouissance* of the maternal body, a body yet undivided, through a system of duality and unsatisfiable desire, against itself.

Not surprisingly, contemporary explorations of the implications of such understandings of subjection and resistance have looked to the disruptive power of body acts and images that, bringing into play parody, multiplicity, and slipperiness, resist a resolution into the fixity of a dichotomous system: Haraway's cyborg body,

Butler's subversive body, Schor's female fetishist, Silverman's exhibitionist. In all these theories, resistance is pinned on the possibilities afforded subjects in a postmodern age of polyvocality, of signifiers floating free of referents, and of simulation. There is a shared sense among these theorists that, as Butler notes,

> just as bodily surfaces are enacted *as* the natural, so these surfaces can become the site of a dissonant and denaturalized performance that reveals the performative status of the natural self. (146)

It is precisely the possibility of this sort of resisting practice that Suskind asks us to consider in *Perfume.* For the novel's central character Grenouille, the power to overcome an abject birth and to live in the world of normal humans is coterminous with his ability to become embodied, through perfume, as any person or thing. Indeed, once acquired, this power opens up to Grenouille the possibility of becoming multiple selves. This sort of overdetermined cultural self is characterized in the novel as:

> a monstrous double creature, like some figure that you can no longer clearly pinpoint because it looks blurred and out of focus, like something at the bottom of a lake beneath the shiver of waves. (182)

In her account of creating her perfume, the body-artist similarly emphasizes the infinite variety of messages and fantasies that perfume is capable of putting into play; by way of imbuing these infinite possibilities with liberatory potential, she contrasts this multiplicity with what are for her the horrors of the totalitarian state:

> if it worked like every fragrance smelled the same on every woman . . . or meant the same to every woman . . . I'd be a billionaire today. It would all be ended. I would have hit the certain chord that every woman would wear this fragrance and it would be all over. It would be like a communist country. One Leader. One Fragrance. One Newspaper. One Book. One Phone Number. And it would be a party line. You have to wait six months to make your one call.

Without referring directly to the disruptive potential of post-modern body practices, even executives in some quarters of the perfume industry now speak of the newest fragrance trend as "the fantasy wave," whose characteristics "must be translated as swiftly as possible" into fantasy scents. Responding to the "massive . . . adversities of life" in the real world these scents "are not one-dimensional . . . they cannot be identified with a single note . . . rather they are complex" (Mensing and Beck 201). Hence, we are currently seeing the proliferation of "celebrity perfumes" such as Elizabeth Taylor's *Passion*, Joan Collins's *Spectacular*, Jaclyn Smith's *California*, Cher's *Uninhibited*, as well as the development of a perfume like Lauren's *Safari*, which promises "A world without boundaries. A world of romance and elegance. A personal adventure and way of life."

To the extent that these perfumes offer consumers the possibility of trying on different personas, they might be considered to have the potential, called for in postmodern theories about the body, to reveal the cultural status of what is taken to be a natural self (Butler, Silverman). In the case of celebrity perfumes, this would seem to be especially true of perfumes such as *Misha* (Baryshnikov) or *Listen* (Herb Alpert). For unlike the perfumes of female celebrities—which conventionally draw upon images of the alluring female constituted in a system of sexual difference—these perfumes evoke worldly attributes of their male creators: "Once in a while there comes a fragrance that hits the perfect note" (*Listen*); "Let it move you like no fragrance ever has" (*Misha*).

The possibility that these perfumes might generate some degree of gender confusion is acknowledged in both cases by the reminder, inscribed at the bottom of these ads, that these are fragrances *for women*. More concretely evoking gender confusion, magazine articles are now beginning to suggest that professional women consider "the crisp, subtle . . . fresh, light" men's fragrances that "wear well in the office" (Gordon 55); men's fragrances are also recommended to women for "scent-sible" workouts at their health clubs (Mark 22).

Both the crossover use of men's fragrances and the proliferation of celebrity perfumes reflect historical changes that have occurred in opportunities opened to particular classes of women and the gender role confusions associated with these changes. If, following Douglas, we read the body as symbolic of the social system, then the use of these perfumes might be considered to reconstitute on the body the social instability, cultural unfixity of the times. Inscribed, through the use of these perfumes, with "the powers and dangers credited to social structures" (Douglas 115), women's bodies might be considered to be playing at the margins of what is constituted as the natural order of things. In this sense, these fragrances seem at least to invite contestation of fixed sex-roles.

Celebrity perfumes in particular seem to open up the possibility of subversion of a body fixed in nature, for while they play to consumers' dreams of becoming "someone else," that someone else is constituted through the media, an image of an original that has no basis in "reality." Thus, in using these fragrances consumers might actually be engaging not *with* a sense of substantive identity but rather *in* a play with and upon the very notion of a substantive identity itself.

A recent *People* piece, "The Great Celeb Sniff-Off," engages directly this parodic potential of celebrity perfumes. In this article, recent popular celebrity perfumes were rated, in order "to help overwhelmed consumers cope." Among the judges convened by *People* was "Sgt. Joe Friday, a German shepherd and trained tracker recently retired from the LAPD"; all the judges were said to be qualified because "all have noses, and all have used, or know people who have used perfume." This article's descriptions of perfumes were similarly parodic, as in this report of judges' perceptions of *Misha*:

> plenty of power and muscle—a rustic musk, multidimensional with long-lasting aromas. Hard-to-please Seinfeld was sent daydreaming about the "inside of a model's sneaker."

Most interesting, as semiotically reinforcing the parodic play here upon fragrance narratives, were the photo representations of the celebrities whose perfumes were judged. These photos showed female celebrities holding relatively small bottles of their perfumes, while the male celebrities were pictured displaying "unnaturally," over-large bottles.[18] Thus ridiculing the constructed body fixed in nature, this play with stand-ins for physiological organs conventionally employed in the constitution of the gendered body simultaneously moved to deconstruct it.

In their ridiculing of gender attributes and interrogation of gender boundaries, in their play with multiple personas and notions of varied unique selves, the above examples point to the possibilities opened up in postmodern consumer culture for the constitution of a postmodern body—multiple, unbound by "natural duality." In this sense, these perfumes and perfume narratives seem to hold out the promise of Suskind's "monstrous double creature," blurring the boundaries of the body, rendering it impossible to pin down. But is this the case? Do they allow for the creation of the monstrous bodies outside of discourses of duality and difference that postmodern theories of multiplicity body forth?

On first reading, it is just this possibility that the denouement of Suskind's novel seems to be argument for. Here, in an ultimate expression of the power that he wields by taking on, through scent, multiple bodies simultaneously, Grenouille indulges in a rite of fragrance self-immolation. Pouring over himself an entire flacon of the powerful scent of multiple budding maidens, Grenouille is immediately torn apart and ingested by desiring hordes. Thus, in what would seem a suitably horrifying postmodern celebration of paradisical multiplicity (Pfeil), Grenouille's multiple subjectivity resolves in disembodiment, the transcendence of a *corporeal* body identity.

But as much as Grenouille's demise might seem to be a symbolic argument in support of the power of a multiple subjectivity to (in this case) implode naturalized body boundaries, to resist subjection through a singular body identity, it is highly problematical, for this novel's conclusion does not offer us the slippery, ever-multiple subject-body that the earlier monstrous double creature passage appears to portend. Rather, it evokes a Christ-like resurrection, and thus asserts in its final pages an essential body-mind dualism.

A close reading of a recent magazine ad for Calvin Klein's *Eternity* perfume exposes a similar tension between the resistive play of multiplicity and blurred distinctions and the desire for resolution into a stable unitary subjectivity. This ad conventionally takes up several full magazine pages and, as in the case of the *KL* ad, uses blacks, whites, and greys to effect a flattening and emptying of its images of any clear representational content. This flattened nonspecificity is reinforced by accessorizing the bodies pictured on the pages with shapeless, soft, colorless clothing *and* sharp-edged futuristic shapes. The semiotic result is a good deal of age, identity, and gender confusion.

For example, in one full page of the ad, three bodies lie in repose: a young boy and two (two different? two of the same?) what appear to be pre-pubescent young women. One of these stares into space and the other lies sleeping; the boy, who is directly behind and above the staring figure, looks down. Unconnected by any mutual gaze, their bodies are rather linked visually, by an elongated pyramidal solid and an open circular hoop. The word ETERNITY appears in the upper left of the page. Three other pages are also inscribed ETERNITY, and picture: The shapes alone; a young girl straddling the pyramid solid and encircled by the hoop; a young girl looking into a frame, with a young boy mirrored back to her gaze. On the two pages of this ad with no text are images of the face of a young boy looking down and another young boy walking, with the hoop.

Combined, these images seem designed to confuse and shake up a sense of fixity in the body—in identity, in desire, in sexuality—and to challenge a system that would insist upon categorizing in difference what are being presented as openly "innocent," that is, unsignified, uninscribed bodies. In this sense, the play with shapes that seems initially to speak to male/female duality appears ironic, as much as iconic.

But if it is ironic play, its effect as challenge is subverted by the iconic image that closes the ad, an image that is the only one among all of them that I have seen being used alone, as a single-page advertisement. This is the image of a mature woman cradling the head of one of the young boys to her breast, and inscribed ETERNITY.

The direct referentiality of this image to motherhood startles as much as the preceding images startle with their refusal of any singular referent. This is not the image of a Kristevan *jouissance* before and outside of signification, as one might conceivably read the images preceding it. It is rather, the image of the *stabat mater*, the maternal body appropriated into the symbolic. Whatever boundary-shattering may have been accomplished by the previous images, this final image puts it all back together again. It enshrines the female body as a culturally constituted mother: loving, self-sacrificing, locked in erotic attachment with her male child.[19]

SEMIOTIC DISRUPTION AND THE CONSTITUTION OF A POSTMODERN BODY

Taken as a cautionary tale, Grenouille's demise at the end of *Perfume* is an argument against a simple politics of resistance through multiplicity, for it reveals how easily a postmodern practice which champions multiplicity against the naturally embodied unitary subject can reinscribe an essential body, one that exists in contrast to "the mind" and within nature. Indeed, in the novel just such an essential body is evoked in Grenouille's claim that "a power stronger than the power of money or the power of terror or the power of death" was nothing if it failed to give him himself, a self that he locates *in* the body:

though his perfume might allow him to appear before the world as a god—if he could not smell himself *and thus never know who he was,* to hell with it, with the world, with himself, with his perfume. (306, italics added)

In a similar fashion, the possibility of a singularly authored body identity, which the body-artist raises in the second narrative-strip with which I opened this essay, is undercut by reference to a natural body, prior to signification. For in continuing, the body-artist's narrative attributes the promise of a unique but changeable identity *not* to some self that the perfume user may select from among infinite multiple selves that perfumes invite her to try on, but to a body that operates biochemically to personalize the olfactory effect of any perfume:

fragrances are changed with the chemistry of your body . . . working with chemicals within your body . . . that's why there's an opening, a middle and a closing note [to a fragrance] because the opening note of the fragrance is one, then the chemistry of your body is starting to take it in, starting to warm up, the heat of your body . . .

Moreover, even as they assert a body in nature, these perfume narratives persist in inscribing it in terms of sexual difference. For example, multiple references throughout *Perfume* to Grenouille's approaching the power "of a god," a section midway through the novel in which Grenouille "goes to the mountain" for an extensive period of fasting and visioning, as well as the final scene of Christ-like resurrection, all suggest that transcendence is an attribute of the marked body—marked with its scent, marked with the phallus. Hence, the above distinction between Grenouille's appearing "before the world as a god" and his inability to smell himself is a distinction between the phallic *poseur* and the phallus embodied as a physical materiality. Similarly, the *Eternity* semiotics reveal how maternity still serves in postmodern culture as the organizing frame for all other features of the female body: Bodies of mothers, daughters, maidens, lovers are at any point in time reducible to the maternal body.

In her recent work on gender identity, Butler contends that one of the major impediments to postmodern deconstructions of the signified body is the pernicious assertion of just such "a discourse of primary and stable identity" (136) as we see in the above examples.

Thus, Butler argues, public body acts that challenge gender identity, such as cross-dressing and drag performance, constitute "local strategies" for exposing the fiction of this internal, a priori identity (149).

Drawn to Butler's vision, I have been titillated, and my optimism has been aroused, by the snippets of postmodern perfume culture which mix up erotic images, ridicule romantic fantasies, tease with the promise of becoming "any body," play with gender—and even species—categories and thus, to employ Butler's phraseology, reveal the "performative status" of the body. But I also share with many postmodern critics—Butler among them—a suspicion that "*transformation* . . . requires more than disruption, uncoupling, slipping away" (Pfeil 76). In particular, I am concerned that our postmodern efforts to deconstruct the body may, to paraphrase one critic, be more a radicalism of gesture than action (Gitlin 358).

This examination of perfume narratives suggests that such concern may be well-founded. For example, it has revealed the power of discourses about bodies in nature (the reproductive body, the libidinal female body) to close gaps that may be exposed through practices or narratives that evoke such bodies, even as they serve to ground efforts to displace them. It has exposed ways in which such practices or claims are engaged in a complicity against the very subjects who employ them.

In some sense, of course, these closures and subversions may be a feature of the specific properties of fragrance itself. As the two narrative-strips with which I began this essay suggest, this is because scent invokes a subjectivity that is configured simultaneously in terms of a universalized corporeal essence—the product of a closed system of signification (nature)—and an internal, personalized self, constituted and reconstituted in an open system of signification (culture).

The first of these two narrative-strips, for example, imbues scent with a *corporeal* concreteness; this same concreteness is expressed by one perfume industry executive as the ability of ". . . fragrances [to] travel directly to the brain without any interpreters" (Green 229). In a popularized scientific account scents are said similarly to "speak directly to . . . [the] primitive network of neurons govern[ing] basic emotional reactions related to survival, such as sexual desire . . ." (Orenstein and Sobel 66).

This notion of scent as intrinsic to processes that are themselves constituted as intrinsic to the natural functions of the body not only makes it difficult to tease out the specificity of the discourses at work constituting the body/scent link, but also undermines efforts to contest these discourses, through performances staged *on the body.* Indeed, in a postmodern environment that construes the body as constituted, such performances seem to sustain the body as a materiality that resists the most scrupulous and rigorous efforts either to dislodge it or to read it as a fluidity of identities.

This is the materiality of a modern body. Constituted through Freudian and biological discourse, this body is inscribed as with an *internal* substantiality. In the face of the deconstruction of such "surface" features as gender attributes—movements, clothing, even body parts—this modern body may actually be taking on an even greater substantiality: the modern "sense of self" positioned as fixed, a priori, in a postmodern environment when all else seems to have been dislodged, disengaged, disembodied.

Yet at the same time scent has an illusory quality. Sensually and symbolically open to contextualization and personalized interpretation, scent evokes responses that are contingent upon individual experience and history. As a recent *Vogue* piece asserts, the sense of smell "can conjure memories from thin air" ("Essence" 274). While it might be "the perfume of a lover, the smell of fallen leaves . . . [that] forcefully evokes feelings" (Orenstein and Sobel 68), these are feelings that belong uniquely to the individual.

Products of the accumulated experiences of an individual's life, the feelings called forth by scent are indisputably cultural, the result of mediations between events and an individual's interpretation of them. They are constituted in an open system of signification. But invoked through the apparatus of the body, they become expressions of an Aristotelian "true self." As the second opening narrative-strip submits, "it's theirs . . . their very own, and no one else can have it."

The narrative collected from a victim of childhood abuse perhaps best exemplifies this:

> I smelled my Mom's boyfriend's cologne on a man in a drugstore years later and after being so afraid I thought I might lose control of my bladder, I was struck with a desire to cut off his penis. I was seized with hatred and fury.

What is especially significant in this narrative is that it is her bladder that the narrator recalls first responding to the fear that the scent calls forth, and that it is the perpetrator's penis that is the first focus of her rage. For in these rememberings is revealed a politics of the body that is not reading *against* a historically constituted body but is rather depending *upon* the coherence, the fixity, the integrity of a body as a ground from which to launch a resistance.

In a theoretical environment that attributes agency to refusal, to multiplicity, to the ability to elude signification, and simultaneously denies the resisting subject unproblematized theoretical ground on which to stand, we should hardly be surprised that the body becomes reconstituted as a sort of theoretical firmament. As Gitlin has recently pointed out, "as the ontological bedrock shakes, nostalgia will return for the old unmoved movers—the unbudgeable signified, the true, the essential, the godly" (358). In closing here with thoughts about the return to body, however, it is not my intention to invoke a materialism as against a postmodernism. Rather, I am concerned to understand postmodernism—position and practice—in the context of the ongoing operation of fairly specific and tangible material relations.

This analysis has illuminated ways in which the structures of late Capitalism and postmodern theory intersect—to open opportunities for resistance *and* to derail and subvert resistance practices. In questioning the power of subversive bodies to reveal the discourses, explode the categories, that subject them, I am not denying the strategic potential of a politics of gesture, image manipulation, refusal. Rather, I am calling for a scrutiny of the consequences for the body that is being engaged at the front line of this politics, and an assessment of what these consequences might hold for us, who so engage it.

NOTES

1. I wish to thank Vassar College, whose support made it possible for me to present an earlier version of this paper at the American Ethnological Society meetings in Atlanta, Georgia, April 1990. I wish also to thank Madeline Cohen and Christina Yoo for their diligent library research, and the students of Women's Studies 130 who helped in the collection of perfume advertisements. Final and special thanks to Barbara Page, who read and commented on an earlier version of this essay, and to Miriam Novelle, who so generously shared with me her experiences in the world of perfume.

2. A similar argument is also being made in the first narrative strip in its assertion that "he who ruled scent ruled the hearts of men," and points to the link, particularly common in fragrance advertising, between a female body that is simultaneously "natural" and capable of multiple "cultural" self-constructions.

3. The proliferation of technologies for instant communication in the postmodern era and the penetration of all media into everyday life mean that the media in postmodern consumer society "are not isolated from each other but refer to each other continuously" (Angus 342). It is through this interchange (rather than in direct reference to any "real" ideal, object or event) that ideas, objects, and events are seen to be constituted as "real" (see also Angus and Jhally). While I find this conceptualization useful for the analysis that I am doing here of magazine and television advertisements of an up-scale product, I use it cautiously. For its tendency to assume ready access to what are frequently characterized as "freely" circulating signs and messages too readily distracts us from analyses of the structures of power that limit both access and what is placed into circulation.

4. Her account of this process celebrated the fact that she made things up as she went along—". . . a hippie out of Woodstock . . . what did I know about doing a major fragrance?"—and so struck me as a bonafide postmodern tale.

5. In his recent review of two novels, Pfeil argues for their being understood as postmodern because of their "distinct . . . strategies for effacing the boundaries between order and nightmare, laughter and horror . . . the curious powerful mixture of delight and discomfort, *jouissance* and revulsion . . . [that] are symptomatic of our own entanglement in an ideological web of themes and discourse we have come to describe as the postmodern" (59). As Suskind employs similar strategies to shake up what would be taken as fixed categories, I have followed Pfeil and claimed *Perfume* as a postmodern work.

6. This phrasing is somewhat awkward, but I have used it in preference to one that would elide narrative and the speaking subject.

7. Following Lovejoy, this narrative attributes the development of bipedalism, a sexual division of labor (in production and reproduction), the incest taboo (in the form of the nuclear family), and such epigamic features as "the conspicuous penis of human males and the prominent and permanently enlarged mammae of human females" (Lovejoy 346; see also Cohen and Mascia-Lees for a critique), to the evolutionary advantage conferred upon early human groups in which males and females were monogamously pair-bonded.

8. It is, in this sense, much like Kristeva's semiotic.

9. Daly and White wrote on "Psychic Reactions to Olfactory Stimuli" in 1930; but the article is cited as recently as 1981. See especially Schleidt, Hold, and Attili.

10. This is surprisingly like the 1865 narrative of Rimmel, in *The Book of Perfumes*: "like the maiden ripening into the matron, the flower

becomes a seed, and its fragrance would forever be lost, had it not been treasured up in its prime by some mysterious art which gives it fresh and lasting life" (3).

11. As Haraway ("Investment Strategies") quite correctly points out, this movement was one that spoke to white middle class women's interests. Indeed, the historical association of black women with unfettered sexuality, and the fact that such an association continued to inform actions and ideologies that contributed to black women's oppression and exploitation, delimited the relevance of "sexual politics" to a very narrow and privileged group.

12. As Susan Bordo continues to point out in her readings of the slender female body, female desire, envisioned as "other" and mysterious, is particularly dangerous in phallocentric cultures, "threatening to erupt and challenge the patriarchal order" (103).

13. See also Baron; Van Toller; King.

14. Similarly, in visualizing her fragrance the body-artist evoked an image of a naked woman, surrounded by flowers "playing in this fur blanket with her fingers and toes through the blanket. And you can feel her hairs and her breasts tingling through the fur." Yet, in conveying this auto-erotic image to perfumers and marketers, she had to translate it into terms of heterosexual desire. As she notes, "How do I walk into a room with fifteen straight looking executives and the president of my company saying, 'I want this fragrance to be like this woman on the blanket, with her clit rubbing against the fur'?"

15. It is relevant that both of these quotes are from women who have achieved unusually high positions in the economic and social structure of postmodern consumer society. This suggests the need in our analyses of consumer practices to attend to class as well as racial and ethnic difference, something which is beyond the scope of this initial exploratory essay.

16. In her feminist reading of contemporary romance novels for women, Modleski offers a similar understanding of the "romantic," asserting that women's "longing to be swept away" (53) coexists with the knowledge that male attraction is something to be won. Modleski's analysis is richly detailed, and its argument that "the desire to be taken by force . . . conceals anxiety about rape and longings for power and revenge . . ." (48) quite usefully challenges earlier psychological models while it establishes a means for examining the political possibilities opened by women's romantic fiction. Special thanks also to Fran Mascia-Lees for first directing my thinking toward contemporary notions of the romantic.

17. In a recent critique of theories linking resistance to the power of the libidinal body, Bartky has argued that if the female body will be the

locus of any resistance, it will be a resistance stimulated by the contradiction between "the relentless and escalating objectification of women's bodies" and the "unprecedented political, economic and sexual self-determination" enjoyed by women (82).

18. The exception is the photo of Cher, who demurely holds a bottle midway in size between the regular bottles held by the other female celebrities and the over-sized bottles displayed by the males. This semiotic exception might relate to the ambiguity of Cher's public sexual image.

19. There is an *Eternity for Men* as well. Early advertisements for this fragrance, in stark contrast to those for its female counterpart, pictured a simple bottle on an otherwise blank page. Recently, however, *Eternity for Men* magazine ads have appeared which substitute a male figure in the place of the adoring mother, complete with boy child held to the chest. This "mothering male" image is being used prodigiously in advertisements for a wide range of products, but space does not allow a further analysis of this provocative dimension of the *stabat mater.*

WORKS CITED

Angus, Ian. "Media Beyond Representation." *Cultural Politics in Contemporary America.* Eds. Ian Angus and Sut Jhally. New York: Routledge, 1989. 333–346.

Angus, Ian, and Sut Jhally. "Introduction." *Cultural Politics in Contemporary America.* Eds. Ian Angus and Sut Jhally. New York: Routledge, 1989a. 1–14.

———. Eds. *Cultural Politics in Contemporary America.* New York: Routledge, 1989b.

Baron, R. A. "Perfume as a Tactic of Impression Management in Social and Organizational Settings." *Perfumery: The Psychology and Biology of Fragrance.* Eds. Steven Van Toller and George H. Dodd. London: Chapman and Hall, 1988. 91–104.

Bartky, Sandra Lee. "Foucault, Femininity, and the Modernization of Patriarchal Power." *Feminism and Foucault.* Eds. Irene Diamond and Lee Quinby. Boston: Northeastern University Press, 1988. 61–86.

Bordo, Susan. "Reading the Slender Body." *Body/Politics: Women and the Discourses of Science.* Eds. Mary Jacobus, Evelyn Fox Keller, and Sally Shuttleworth. New York: Routledge, 1990. 83–112.

Butler, Judith. *Gender Trouble: Feminism and the Subversion of Identity.* New York: Routledge, 1990.

Byrne-Quinn, J. "Perfume, People, Perceptions and Products." *Perfumery: The Psychology and Biology of Fragrance.* Eds. Steven Van Toller and George H. Dodd. London: Chapman and Hall, 1988. 205–216.

Cixous, Hélène. "The Laugh of the Medusa." *Signs* 1 (Summer 1976): 875–899.

Cohen, Colleen B. and Frances E. Mascia-Lees. "Lasers in the Jungle: Reconfiguring Questions of Human and Non-Human Primate Sexuality." *Medical Anthropology* 11 (1989): 351–366.

Daly, C. D. and R. Senior White. "Psychic Reactions to Olfactory Stimuli." *British Journal of Medical Psychology* 10 (1930): 70–87.

Douglas, Mary. *Purity and Danger.* London: Routledge and Kegan Paul, 1966.

"Essence: A Love Story." *Vogue Magazine* (May 1990): 272–279.

Firestone, Shulamith. *The Dialectic of Sex.* New York: Morrow Quill, 1970.

Foucault, Michel. *The History of Sexuality, I: An Introduction.* Trans. R. Hurley. New York: Vintage/Random House, 1980.

Gitlin, Todd. "Postmodernism: Roots and Politics." *Cultural Politics in Contemporary America.* Eds. Ian Angus and Sut Jhally. New York: Routledge, 1989. 347–360.

Gordon, Jill. "Your Travel Case: The Sweet Smell of Success." *Vis à Vis* (November 1988): 55.

———. "The Great Celeb Sniff-Off." *People Magazine* (19 February 1990): 52–55.

Green, A. "Fragrance Education and the Psychology of Smell." *Perfumery: The Psychology and Biology of Fragrance.* Eds. Steven Van Toller and George H. Dodd. London: Chapman and Hall, 1988. 227–233.

Haraway, Donna J. "Investment Strategies for the Evolving Portfolio of Primate Females." *Body/Politics: Women and the Discourses of Science.* Eds. Mary Jacobus, Evelyn Fox Keller, and Sally Shuttleworth. New York: Routledge, 1990. 139–162.

———. "A Manifesto for Cyborgs: Science, Technology, and Socialist Feminism in the 1980s." *Socialist Review* 15 (1985): 65–108.

———. *Primate Visions: Gender, Race and Nature in the World of Modern Science.* New York: Routledge, 1989.

"The Hidden Power of Body Odors." *Time Magazine* (28 January 1986): 67.

Jameson, Fredric. "Postmodernism and Consumer Society." *The Anti-Aesthetic: Essays on Postmodern Culture.* Ed. Hal Foster. Port Townsend, Washington: Bay, 1983. 111–125.

King, J. R. "Anxiety Reduction Using Fragrances." *Perfumery: The Psychology and Biology of Fragrance.* Eds. Steven Van Toller and George H. Dodd. London: Chapman and Hall, 1988. 147–184.

Kristeva, Julia. *Desire in Language.* New York: Columbia University Press, 1980.

——. "Stabat Mater." *The Kristeva Reader.* Ed. Toril Moi. New York: Columbia University Press, 1986. 160–186.

Laqueur, Thomas. "Orgasm, Generation, and the Politics of Reproductive Biology." *The Making of the Modern Body.* Eds. Catherine Gallagher and Thomas Laqueur. Berkeley: University of California Press, 1987. 1–41.

Lovejoy, C. Owen. "The Origins of Man." *Science* 211 (1981): 341–350.

Mark, Lois Alter. "Skin, Scent and Hair." *American Health* (8 November 1989): 19–21.

Mascia-Lees, Frances E., Patricia Sharpe, and Colleen Ballerino Cohen. "The Female Body in Postmodern Consumer Culture: A Study of Subjection and Agency." *Phoebe* 2.2 (Fall 1990): 29–50.

Mensing, J. and C. Beck. "The Psychology of Fragrance Selection." *Perfumery: The Psychology and Biology of Fragrance.* Eds. Steven Van Toller and George H. Dodd. London: Chapman and Hall, 1988. 185–204.

Modleski, Tania. *Loving with a Vengeance: Mass Produced Fantasies for Women.* New York: Routledge, 1982.

Orenstein, Robert. "The Scent of Love." *Cosmopolitan* 207 (August 1989): 203–205.

Orenstein, Robert and David Sobel. *Healthy Pleasures.* Reading, Massachusetts: Addison-Wesley, 1989.

Pfeil, Fred. "Potholders and Subincisions: On *The Businessman, Fiskadoro* and Postmodern Paradise." *Postmodernism and Its Discontents.* Ed. E. Ann Kaplan. London: Verso, 1988. 59–78.

Pratt, Jonathan. "Notes on the Unconscious Significance of Perfume." *International Journal of Psycho-Analysis* 23 (1942): 80–83.

Resnick, Jill. "Suitable Scents: A User's Guide to Fragrance at the Office." *Working Woman* (December 1989): 120–122.

Rimmel, Eugene. *The Book of Perfumes*. London: Chapman and Hall, 1865.

Schleit, Margaret, Barbara Hold, and Grazia Attili. "A Cross-Cultural Study on the Attitude Towards Personal Odors." *Journal of Chemical Ecology* 7 (1981): 19–31.

Schor, Naomi. "Female Fetishism: The Case of George Sand." *The Female Body in Western Culture*. Ed. Susan Suleiman. Cambridge: Harvard University Press, 1986. 363–372.

Silverman, Kaja. "Fassbinder and Lacan: A Reconsideration of Gaze, Look and Image." *Camera Obscura* (19 January 1989): 55–84.

Smith, Paul. *Discerning the Subject*. Minneapolis: University of Minnesota Press, 1988.

Stoddard, D. M. "Human Odor Culture: A Zoological Perspective." *Perfumery: The Psychology and Biology of Fragrance*. Eds. Steven Van Toller and George H. Dodd. London: Chapman and Hall, 1988. 3–18.

Suskind, Patrick. *Perfume: The Story of a Murderer*. Trans. J. E. Woods. New York: Pocket Books/Simon and Schuster, 1986.

Van Toller, S. "Emotion and the Brain." *Perfumery: The Psychology and Biology of Fragrance*. Eds. Steven Van Toller and George H. Dodd. London: Chapman and Hall, 1988. 121–146.

Wilson, E. O. *On Human Nature*. Cambridge: Harvard University Press, 1978.

Wittig, Monique. "The Mark of Gender." *The Poetics of Gender*. Ed. Nancy K. Miller. New York: Columbia University Press, 1986. 61–73.

5

Vandalized Vanity: Feminine Physiques Betrayed and Portrayed

INTRODUCTION

This paper documents the power of culture to contour, inscribe, embellish, and mortify the female form. The body is regarded here as a double entendre of embodiment: it personifies itself through denaturalization and consequent renaturalization. In no area is the pretense of the "cultural as natural" so strongly articulated than in the construct of the body. The relations of the "natural body" will be questioned through analyses of the diverse ways in which the female body is written as the "cultural phenotype" (Turner 8) of broader social constructs of masculinity and femininity, passion and restraint, strength/power and weakness.

My purpose is twofold. The physical self, regarded as the embodiment of systems ordering male and female relations, hierarchy and power, will first be placed in historical and theoretical context.

Second, data collected ethnographically with competitive women bodybuilders is offered as case study material highlighting the body as an interpretive discourse on Western gender. The potency of the symbolic in altering the form of the body through dieting, exercise, and adornment is woven throughout this discussion. Biology emerges as a socially mediated category in which culture imperso-nates nature and profoundly affects the meaning of female, woman, and femininity.

Despite early interest by Darwin (1872), and subsequent de-velopments in the field of kinesics (e.g., Hall; Fischer; Messing), symbolic and interpretive anthropology (e.g., Hanna; Martin; McDougall; and Postal) and discourse analysis (e.g., Clifford; Foucault; Marcus; Hanks), the body as an explicit domain of re-search has been relatively ". . . absent in social theory . . . and has merely made a cryptic appearance" (Turner 2,6) as a field worthy of intellectual endeavor (for exceptions, see for example, Polhemus; Armstrong; Blacking). A very recent impetus for the study of the body has come from postmodern theory which has contributed to a prolific procreation of publications on the subject. In the field of anthropology, feminist thinking conjoined with postmodern the-orizing has suggested provocative new paradigms from which to encounter the physical self, rendering the body in greater con-textuality, complexity and as more reflexive than in previous work (e.g., Mascia-Lees, Sharpe, and Cohen 1989 and 1990; Flax; Tress).

The body has emerged in some contemporary anthropological thinking as "colonized" by cultural rules, as textualized by power, and as subject (in terms of body image) and object (as the self con-ceives of and responds to the soma).[1] In this way the body as a natural symbol is conterminously intertwined with its function as a cultural symbol.[2] The discussion that follows represents a synthesis of these views that is informed by feminism, postmodernism and interpretive anthropology.

THE CULTURE OF BEAUTY

It may be argued that Western cultures are keen to elaborate sex differences, and consequently that feminism's perceived threat doesn't lie in changing sex roles per se, but rather in the potential for altering somatic and sartorial insignias of sex. This Western bio-centric view of gender implies not only that there are behavioral and personality correlates for the sexes, but that bodily characteristics (apart from the genitalia) such as muscularity/fatness, hardness/softness, and triangular/hourglass contours are sexually dimorphic and even bipolar. I am not suggesting that there are no differences in body fat and muscle between the sexes, but that these attributes have a strong cultural dimension. And it appears that this dimension is saturated with metaphors of the opposition between the sexes. Such a process results in "incorrigible propositions" about gender (Kessler and McKenna 5) that thrive on regarding the sexes as bimodal and exclusive rather than a continuum. According to Gallagher and Laqueur (viii) the oppositional view of the sexes is a relatively recent phenomenon; originating in the eighteenth century. Prior to this period, the prevailing gender theory was one that stressed the homologies of male and female.

This radical new gender schema based on a cultural ideology of women's physiologies has informed social meaning profoundly. Thus, women are labeled the weaker sex ". . . and then . . . [it is] . . . assume[d] nature designed her that way" (Freedman 57). That women are the weaker sex is a central theme in the contemporary construction of the female body. This is demonstrated in the curvaceous, rounded, soft forms that display a decided difference in visible muscle between women and men in popular media images. However, women's body norms include alternatives, as hourglass configurations are contrasted with tubular or athletic contours. The postmodern shape of women includes a whole range of types. In one corner is the form of the denied and restrained female body, similar to the virtually unattainable image of an adolescent boy superscripted with large breasts and epitomized in the *Playboy* bunny centerfold or media stars such as Vanna White. In another corner is the competitive female bodybuilder with hypertrophied muscles, stamina, strength, and power. Is this contour a transformation of the former; a contestation of male privilege—another kind of power suit? Or is the female bodybuilders' physique another message in gender fixity? Placing the female body in historical perspective will facilitate the exploration of such questions.

What crosscuts the variety of female physiques is the linkage of beauty with femininity. Historically, Western somatic and sartorial ideals articulate with a culture of beauty. The history of women's bodies in the United States is indeed a history of beauty.[3] While beauty has been amply noted in the theorizing of biological anthropologists concerned with the role of physical attractiveness in human evolution (e.g., Symons), only recently has beauty and attractiveness been explored by social scientists as in the work of Freedman and Banner among others.[4]

The culture of beauty chronicles the modification and mortification of the somatic self. It is intimately tied as a symbolic structure to fashion, adornment (Banner 17–45), dieting, and exercise and as an economic structure to capitalism. Women have come to be defined primarily in terms of allure and adornment, highly visible displays, and demonstrations of men's success (Freedman 106). Capitalism and beauty conspire as a system in which seasonal changes in fashions as well as cycles in the female contour sell products. Women's ambivalence about body concept is institutionalized as a mechanism in the economics of beauty. Because women are valued for how they look, not what they do, they are more affected by ephemeral attractiveness norms than men (Freund 85). "When beauty images change, bodies are expected to change as well, for nature cannot satisfy culture's ideal" (Freedman 6).

Culture's ideal for the American woman incorporates two essential attributes that buttress the beauty norms: youth and femininity. Freedman (195) proposes that beauty embodiments foster neotonic traits in women, while male attractiveness norms emphasize dominance traits. Cultivating childlike qualities and images as a beauty strategy undermines women's power and reinforces the social dominance of males. It may also explain historically why women's soma and sartorial system have emphasized petiteness/smallness as well as the central contradiction in the sport of competitive women's bodybuilding around the issue of muscular size and femininity. In summary, femininity is clearly connected to attractiveness producing a gendered trait complex in which beauty, neotony, and femininity intersect and are culturally constructed as incompatible with the traits of masculinity such as power, strength, instrumentality, and adulthood (see Broverman et al. 66).

HISTORY: THE BODY BOUND AND UNBOUND

Chronologically, fashion has proven to be an arena for social struggles, as has the body (Ewen and Ewen 150). "Each new wave of feminism includes some aspect of dress reform, for a woman's role is costumed and dictated by her robes" (Freedman 86). Dress reform is necessarily body reform, either directly by confronting the constraints in clothing or indirectly by offering alternative models of the female contour such as unrestricted or even athletic options. The liberation of the female soma and sartorial system is intertwined with the conjoining of feminism, health, and fitness movements historically.

A direct relationship occurs between the metaphor of the weaker sex as a social category and its transformation into physical weakness. Conversely, athletic skill and strength are associated with social efforts towards women's equality. As a result, tension is created between femininity/weakness and athleticism/strength. Women must be ever watchful about crossing the beauty barrier into the physical domain of men; muscularity and strength are associated with male athletic prowess. Asymmetry in beliefs about such alleged gender dichotomies in terms of morphology and concomitant strength are reflected in and accelerated by fashions and ideal body types. These asymmetries transform quantitative gender dimorphisms into renaturalized qualitative differences rendering beauty as biocentric and woman as "the other."

The eighteenth and nineteenth centuries in America may be regarded as the age of the corset. This form of female sartorial restraint may be traced to fifteenth century Europe where the bodice evolved into the busc, a garment reinforced with a variety of inflexible materials such as wood, metal, and horn eventually evolving into the corset (Brain 84; O'Hara 76; Bullough and Bullough 184–185). The corset created the hourglass ideal by pushing tissues down into the gluteal region as well as pushing up the breasts (Brain 136). Banner (47–50) has classified this image as the "steel engraving lady," a fragile, sylphlike vision of womanhood characterized by an eighteen inch waist (the product of tightlacing or the physical act of putting on the corset and cinching it), rounded hips, and tiny feet. The female body was circumscribed and reformed throughout the lifecourse to conform to this ideal. After birth a belt was tied around the baby girl's waist, a process which permanently altered the female form. Later, training corsets were worn by school age children (Brain 83; Bullough and Bullough 186) and pregnancy corsets were even designed to accommodate the pregnant and nursing woman.

The hourglass embodiment, along with the custom of tightlacing, necessarily inhibited women's consumption of food. Consequently, the genteel lady was never to appear hungry publicly (Banner 47–50), a custom of self-denial still promoted today. The message of feminine denial of the appetites contradicted the images of sexual and reproductive profusion evident in the roundness of hips and breasts. The use of the corset was a ritual of mortification, for it represented the passions as vulgarity, denial as frailty and femininity (Ewen and Ewen 131–141).

Notions of the delicacy of womanhood included the prevalent

view that the female lifecycle was inherently draining. Indeed, tight-lacing became a self-fulfilling prophecy, for it produced uterine and spinal problems begetting the fragile and wan woman the medical community of this period expected. Social norms denoted this body text as well, for it became "fashionable to be ill, pale and thin" (Banner 47–51, 128). In this regard Mary Douglas (1966 in Douglas 1973: 98) has noted ". . . the experience of consonance in layer after layer of experience and context after context is satisfying. I have argued . . . there are pressures to create consonance between the perception of social and physiological levels of experience." The steel engraving lady illustrates Douglas's principle of consonance. Just as today, fat could nullify beauty, for the large and overweight woman of the period was one who had given in to her appetites. The paragon of femininity was ethereal beauty, one whose passions were as whittled as her waist.

The latter half of the nineteenth century produced far greater variety and extremes in sartorial profiles than the former century and a half. By the 1870s the Grecian bend, an "s"-like silhouette produced by tightlacing and bustles, became the unparalleled ideal. Hoops over the gluteal area produced a tableau of lordosis. This contour of sexual receptivity continued the sharp contrast between body images and the Victorian denial of desire and voracity (Banner 97; Ewen and Ewen 140, 142).

The mid-nineteenth century also engendered two important movements, women's rights and health. Both modified the configuration of the female contour through efforts at dress reform. Tight-lacing came under attack for the medical problems it caused. Amelia Bloomer, an antagonist of the corset, designed a freer alternative: a short dress worn over Turkish trousers, a risky design because trousers were associated with men's clothing. *Harper's Magazine* was concerned that women in bloomers would renounce their power over men by becoming their rivals (Banner 83; Ewen and Ewen 154–155). One can imagine that at the time donning bloomers was tantamount to having a sex change today.

With women's rights and dress reform, came a message about body reform as well. While Elizabeth Cady Stanton suggested that exercise and clean living were important for women to have a "healthy glow," Susan B. Anthony maintained that women needed strong bodies as well as minds in the fight for equality (Freedman 229, 164). The fragile, steel engraving lady was challenged by this athletic figure. Here was a physical configuration that suggested strength, a more natural and wider waist with stronger arms, a body ideal persisting throughout the latter half of the 1800s coexisting with the Grecian Bend ideal.

By the last two decades of the 1800s the impact of health reform and feminism were beginning to be felt in the beauty norms for women. Both Banner (106–107, 139) and Freedman (147) report on a new plump model of womanhood that became *de rigueur.* Plumpness, as an exaggeration of the voluptuous ideal, emerged briefly as a sign of health and of beauty. By 1885, the American Association for the Advancement of Physical Education was formed, but women remained ambivalent about exercise, for it was believed that it produced muscular bodies, destroying women's "natural curves" and de facto their femininity.

Despite a seeming burgeoning of athletic interest through the latter decades of the 1800s that resulted in concomitant dress reform for women's sports, the feminine ideal of curves and roundness still held sway. Just as today, advocates of exercise legitimized their position by bracketing the dominant model. Athletic women were touted as voluptuous and in a late nineteenth-century defense it was said "no matter if a woman's biceps may grow hard as iron and her wrists firm as steel, the member remains so softly rounded, so tenderly curved" (Banner 141–143).

By the nineteenth century, the Gibson Girl arrived on the scene as a portent of the early twentieth-century feminine ideal of naturalness. The Gibson Girl was a healthy, strong, athletic, albeit corsetted ideal. I would liken her to bodybuilding's Rachel McLish, an athletic form that pushes ideal body boundaries yet does not violate the essence of "femininity." The Gibson Girl, like the early women bodybuilders such as McLish, offered something new that stretched notions of the ideal feminine form but did not violate its essentialism.

The early twentieth century may be characterized by this new "natural" look for women, an even more athletic contour than the Gibson Girl that overrode the previous short lived bulge in history for voluptuousness in the 1880s. Beauty norms of the twentieth century evolved to incorporate this new "healthy look," replacing the ill health and delicacy implicit in the steel engraving lady. But the new natural beauty was above all a thin figure not a robust one, continuing the established trajectory of denial of desire. The alleged natural look of the 1900s was not achieved without dieting which came into its own at this time (Bordo 83). In this regard Flugel, has suggested that the twentieth century focus of beauty has shifted from clothing to bodies, a metamorphosis entailing in addition to diet, exercise, and even surgery.

The "thinning of America" and the "tyranny of slenderness" (Chernin) are clearly textualized in the image of the flapper of the 1920s. The flapper signaled a new woman who was valued for her independence and nonreproductive sexuality. Her appearance was boyish; her hips were thin, her breasts were bound and her hair was bobbed, symbolic of her youthfulness and freedom. The fashions of this era were emblematic of sports and athletics, and it was during this period that women athletes gained notoriety as public figures (Freedman 2, 149; Banner 279).

The 1930s brought back the buxomness of previous eras, but continued to textualize masculine independence and assertiveness with the advent of square and padded shoulders. Youth as a concomitant of beauty was supplanted by maturity, as a more sophisticated trend for women asserted itself through World War II. The theme of adult womanhood continued even through the 1950s when an alternative body harkening back to the hourglass ideal reasserted itself (Lurie 77–79). By the 1950s female body norms included the mammary mystique, a return to voluptuousness that incorporated childlikeness. Monroe encapsulated this ideal which included sculpting the female physique with girdles, corsets, and brassieres (Guthrie; Banner 284). The athletic figure of the 1920s was not incorporated into the 1950s ideal of curvaceous womanhood. The decline of the athletic body was reiterated in the coterminous decline in funding for women's sports (Banner 286).

By the 1960s and 1970s, feminism brought massive sartorial and somatic changes. A new slenderness emerged embodied by Twiggy, displaying vestiges of the flapper's linearity, but far more extremely. Beauty ideals centered on thinness during this period, and, in the extreme, textualized abandon, emancipation, and nonreproductive sexuality. While women's desire was encoded through a nonreproductive body, one that de-emphasized breasts and hips in contrast to the 1950s body text, the restraint of desire was, paradoxically, represented in the denial of the appetites necessary to achieve the exaggerated slenderness.

This brief overview has focused on dominant themes displayed primarily through public and widely available images of women throughout history. The theme of restraint and power is evident through the culture of beauty, as the idea of women as display underwrote the male privilege to "look." Such profound inequities in status have and continue to be contested in women's efforts to control their bodies and passions. Yet paradoxically, this internalized effort at self-imposed control of the passions and desire is embedded in the larger external control of women manifested through sartorial constraints such as corsets and girdles. Yet women's bodies have also been a site for contesting male privilege as efforts at emancipation have been inscribed in bodies and sartorial shapes (e.g., the Gibson Girl, the flapper of the 1920s, the padded shoulders and triangular shapes of the 1930s women, along with the more mature body type that lasted through the 1950s). The latter body gave way to a liberated innocence expressed in the 1960s ethos of youth. The sartorial text of baby doll sacks, shifts and empire dresses overlay an extreme slender ideal, a narrative of freedom from reproduction that clearly articulated feminist ideals with concomitant technologies of "birth control." The apparent contestation for these emancipated female forms, however, has revealed a scaffolding of male privilege, power and position asserted through the culture of beauty and the continual reassertion of the "feminine" paradigm of frailty, weakness and neotony.

The 1980s have brought alternative models of beauty. While the slenderness of the 1960s has continued its tyranny, a world-wide movement toward health and fitness has resulted in a more rigorously acquired body ideal that includes toned muscles and taut physiques. Gallup Surveys for *American Health Magazine* estimate over 50 percent of Americans get some kind of regular exercise (Raskin 31). The athletic form coexists with a representation of delicacy and voluptuous contours, synchronized with the reemergence of corsets, lace, and the political conservatism of the 1980s. The slimming of America, begun in the early twentieth century, endures as a core feature of feminine beauty summarized by the Duchess of Windsor: "you can't be too rich or too thin." But women can take the athletic imperative too far and can have too much muscle. The remainder of this discussion will focus on the competitive women bodybuilder whose body configurations represent at once the polysemic nature of cultural texts, the blending and blurring of insignias, and the betrayal and portrayal of a biocentric schema of gender whose social construction imputes irreducible differences. The woman bodybuilder embodies oppositions that are buttressed by patriarchy.

COMPETITIVE WOMEN BODYBUILDERS

Within the year of Cory Eversen's crowning (actually she was medaled in the Olympic tradition) for the fifth consecutive time as

1988s Ms. Olympia, bodybuilding's ultimate title, *Cosmopolitan* proclaimed:

> The return of the female form. Stop dieting to distraction and exercising to exhaustion. It's time you learned to *love* your curves (like he always has!). There is something alluring about a bit of hip, an hourglass shape. It's very, well, touchable. . . Try yoga for stretching, toning. You can stay in beautiful shape without muscle building. . . Curves à la Monroe (if she'd worked out a bit more) are what's red hot right now. ("The Return of the Female Form" 182–185)

While these curves are constrained by the resurgence of corsetry, they are equally carved by ideology. Beauty norms not only require conscription by clothing, an outside-inside transformation, but the converse, as the body is cast from inside-outside in terms of diet and metabolic manipulation through exercise. Both soma and sartorial systems encode capitalism and women's worth as artifacts of display. The twentieth century, particularly the last thirty years, has witnessed the emergence of diet and exercise and modern correlates of the nineteenth-century corset.

Bodies, like clothing, manifest a seasonality as testified by *Cosmopolitan.* Not only does clothing shape the physique, but the physique shapes the clothing. Contemporary women with lean and linear muscular contours, for example, do not require corsets, but voluptuous women can be convinced to buy them. Bodies are not only the text of women's rights and roles, but also of capitalism as particular contours sell clothing and cosmetics. Examples abound: pastel powder and lipstick for the delicate corsetted form, and sweatproof tan colors for the athlete or would-be athlete for that healthy outdoor look. Services feed physiques in a host of enterprises such as fat farms and diet centers, bodywraps and anti-cellulite massages. Trim is not enough, the anatomy ante has been upped to include tight and toned, even for the curvaceous woman who is likened to Marilyn Monroe if only she had exercised a little more.

Throughout history the athletic image has made its appearance as an unconstricted and linear form coexisting or even, as in the 1920s, supplanting the dominant hour-glass contour, nevertheless at no time has the muscled woman been regarded as a paragon of beauty. Muscles and femininity are representationally exclusive categories. While athletic bodies have become increasingly popular in advertising and do display a new found muscularity, it is more within the confines of the "toned" as opposed to brawny physique. Women whose muscles reach a degree of size indicative of underlying strength are in fact denoting a marginal body type.

The ethos of the "weaker sex" still prevails in bodily inscription. Muscles are emblematic of the domain of males not females, implying strength and power, socially, economically and physically, a redundancy in symbolic channels. Muscles encode agonic power, the power of male strength, while women's power is hedonic, that of display (Freedman 73–74). Competitive women bodybuilders are therefore embodying status strain or incongruity, and are threatening to a femininity conceived in terms of frailty. While the bodybuilder's muscles seemingly contest agonic power as solely a male prerogative, the sport of bodybuilding is an engram of hedonic display, in which conventional images and meanings of womanhood abound hidden within a new context.

Until recently the athletic ideal was the antithesis of the feminine beauty ideal. The athletic body gained momentum as a result of the dovetailing of two movements: that of feminism in the 1960s and the fitness movement of the 1980s. The fit female form is a toned one with low body fat indicative of diet and of exercise as well. This new soma inscribes women's greater strength socially and economically through the body as text. Today it is not uncommon to see this parodied in pictures of women models flexing their biceps to sell consumer products (Duff and Hong 374). "Jock chic" has come of age for the white middle and upper classes (Freedman 166). Because jock chic is a look for women (for the male gaze), it is by default correlated with the construct of beauty, not necessarily achievement. Its semantic frame still inscribes femininity, for it incorporates the denial of desire and restraint while ironically displaying it. For example, women's bodybuilding, while representing the extreme end of jock chic, has been advertised as television's "sexiest sport" (Duff and Hong 380).

The athletic ideal is indeed disruptive, and consequently cultural efforts at deactivating its potential threat embed and constrain it. Because the body is regarded as natural, female athletic bodies challenge the Western gender schema and its biology of difference, a paradigm of patriarchy and inequality. Paradoxically the athletic body conterminously embodies beauty yet endorses the acquisition of power in what remains a last bastion of male strength, the physical arena. In an unpublished manuscript, Holly Devor has traced the frequency of popular media coverage of women's weight training and bodybuilding and reports of rape. According to Devor ". . . as women became aware of the level of violence, they responded with an interest in developing physical strength" (personal communication). Gaining strength is a way of establishing agonic power in the face of the continued victimization of women. Gaining strength is a rupturing antidote to crimes of power and position.

The athletic ideal has set the stage for a trajectory that some women bodybuilders have pushed well beyond normative beauty standards of toned, taut, and feminine. Images of competitive women bodybuilders' physiques have spread through popular culture via movies such as the now classic *Pumping Iron II: The Women,* talk shows, athletic broadcasting, magazines, and media stars such as Rachel McLish and Cory Everson with her syndicated bodyshaping show. I have relied on a variety of genres in interpreting competitive women's bodybuilding including magazines for bodybuilders which report on contests, organization news, training, and nutrition as well as which provide profiles and gossip columns on the sport and its participants. In addition, I have engaged in participant-observation with women and men bodybuilders for two and a half years in gyms, at contests and at other bodybuilder social events. My experience includes formal and informal interviewing, training as a judge and test judging contests as well as competing as a bodybuilder. I have combined empirical research with humanistic interpretations, relying on an eclectic array of narrative sources.

Heralded as women's most nontraditional sport, competitive bodybuilding is saturated with contradiction. While women have indeed gained strength in a variety of domains, these gains are mediated by the culture of beauty and the potency of femininity. From its infancy in 1979, women's competitive bodybuilding has been arguing about images. Metaphorically, the question may be put as "are we to view these women as beauty or the beast"? This sport is unique as one in which contestants in pursuit of becoming the best may ultimately be ranked lower as a competitor. A woman bodybuilder, unlike a man, may be too good; she may have too much muscle. Women's bodybuilding has wrestled with its three major criteria in judging: muscularity (including "cuts" [argot for depth and leanness], striations and vascularity as well as muscle shape and size), symmetry (proportion) and posing (performing a series of mandatory poses as well as a personalized posing routine). Judges may rank a contestant lower because she is "too muscular" and her physique challenges conceptions of beauty and femininity. Beauty is a very important part of marketing the sport and hence is closely tied to the criteria for the selection of contest winners.

In the last ten years, the competitors have pushed the perimeters on muscular size, and the question of how far is too far before one's womanhood is forfeited continues. This issue marked the first Olympia for women in 1980 and has permeated the most recent Ms. Olympia events. The editorial staff of *Muscle and Fitness Magazine* ("Preview: Ms. Olympia November 25, 1989") accused women's bodybuilding of being ". . . unable to set certain standards for scoring and defining itself. Many purists wish to see it stay as a muscle cult . . . while others wish to see it further glamorized and showbizzed in a bid to win further media acceptance" (76, 199).

As women bodybuilders become more muscular every year, they are approximating a stigmatized or non-normative model of womanhood, yet one that paradoxically embraces dominant themes in the culture of beauty. The ideal for the competitive bodybuilder is pro-normative; they've taken a good thing, athletics and toning, and have been accused of taking it too far. This concerns the sport industry of bodybuilding, consisting of professional organizations, products, services and marketing of the former. The promoters of bodybuilding desire legitimacy and respectability. As big business, the industry in seeking mass appeal is de facto influencing judging criteria and the selection of winners.

The question becomes how far can women pursue muscularity and still sell the sport? National level competitors are well beyond a toned athletic physique. The competitive ideal of Cory Eversen, media personality and spokesperson for the sport, in peak condition is not an image the public finds appealing. In a recent survey that I conducted, college students were shown a contest photograph of Cory Everson and asked to rate her attractiveness and femininity on a scale of 1 (least attractive, least feminine) to 7 (most attractive, most feminine). Out of one hundred fifteen female responses, the average attractiveness rating was 2.9 while the average femininity rating was 2.6, below the median point of the scale. The male responses (n = 43) were similar; the average attractiveness rating was 3.1, while the average femininity rating was also 3.1. While not generalizable, the homogeneity of response bears looking at. In addition, answers to an open ended question regarding what first came to mind when viewing the photograph included terms and phrases such as "disgusting, too big, beastly, looks like a man, intimidating, unfeminine." Yet Cory Everson represents the pinnacle of competitive perfection; she is hailed as the ideal combination of muscularity and femininity by the bodybuilding community at large.

The somatic style of the winners of major contests such as Cory Everson is crucial to the promoters and vendors of the sport whose market extends far beyond the elite group of competitors and includes the fitness needs of middle America. Articles encouraging women to work out seek to legitimize the sport and neutralize the association of weights and workouts as masculine. These invariably include the disclaimer "... any muscle tissue a woman ... might develop will show up on her body as feminine curves" (Weider and Weider 6), a voice from the latter half of the nineteenth century heard once again. Joe and Betty Weider reveal that "the bodybuilding lifestyle ... results in buoyant health which translates into radiant beauty," and lest one doubt the heterosexism in beauty, when women and men train together "they usually find some excellent social opportunities" (111). Neutralization involves calling upon mandates from the culture of beauty.

The culture of beauty is also invoked on a subtler level in judging as well as media promotion. The judges are accused of selecting athletic, slim, graceful, and pretty women whose muscles don't show unless flexed (Gaines and Butler 69), epitomized by Rachel McLish. While the muscle size has increased dramatically over the years, the essence of this commentary lies in the potency of prettiness. The 1988 controversy over the placement of Ellen Van Maris is indicative. In the 1988 Ms. Olympia she placed fifth, a substantial drop from her second place finish in 1987. Van Maris had apparently no flaws other than that the judges thought she was just too muscular (Tanny 90, 92). However, the judges have been criticized for selecting top winners on the basis of their beauty, and it has been argued that Van Maris's placement reflects her "lack of marketability" (i.e., attractiveness) (Annand 18), rather than her standing as a bodybuilder.

Bodybuilding is not far removed from the elements of the beauty contest. This conclusion is supported by a content analysis of adjectives used in profiles of twenty-eight women and men competitors from *Flex, Iron Man,* and *Muscle and Fitness.* In eleven out of fourteen randomly selected articles on women, adjectives describing attractiveness were cited in addition to descriptions of their physiques, but in fourteen articles about men, not one referred to the physical attractiveness of the competitor.

Beauty in the female body is textualized as display, sexuality, youth, and even blondness. The latter trait is included as one of the neotonic attributes of womanhood. "While one in four American women is born blond, only 5 percent remain so naturally after puberty . . . so staying blond longer means looking younger" (Freedman 196). Women's bodybuilding includes these elements in its media images. Vedral in describing the 1988 Ms. Olympia states: "The moment they stepped out on stage I noticed something different. They were pretty. They were sensual. They were appealing" (124). (I'm tempted to add they must have looked like the Stepford Wives!) Vedral explains that "it seems to be a combination of not being as big and bulky, having more symmetry and paying more attention to grooming such as hair, makeup and nails" (127). Finally, out of the top nine winners of the 1988 Ms. Olympia, seven were blonde, an insignia of neotony, a rendering of the culture of beauty.

CONSTRUCTING FEMININITY: THE ISSUE OF NATURALNESS

The contradiction "of beauty or the beast" in the competitive body is clouded by the use of anabolic steroids, derivatives of testosterone which allegedly cause muscle growth and strength increase without androgenic side effects (Haupt and Rovere 469). In in-depth interviews with nine women bodybuilders who were questioned on whether women could develop muscles that were too big, there was 100 percent agreement that it is not possible without the use of steroid drugs. Some individuals maintain that genetics may limit overall potential for development.

The argument behind the apparent contradiction in judging criteria includes the issue of steroids as a factor in the muscularity and femininity debate. In the bodybuilding community, it is common knowledge that women get "freaky huge" with the use of steroids (Duff and Hong 524). It is in this arena that personal theories regarding steroids and their consequences abound. Drug testing for women was initiated in 1985 in efforts to provide a more acceptable image for the public at large. This was approximately five years prior to testing the competitive men. In a 1986 interview regarding drug testing for women, Jim Manion, President of the National Physique Committee (an association of amateur bodybuilders), stated:

> I was one of the people who pushed for more muscle in the girls . . . the women's federation at the time picked winners that were too soft. . . But then it seemed that the women went

> too far. . . [They] were looking too offensive for the net-
> works . . . drug testing became an imperative. . . I personally
> don't feel men need to be tested, because you're talking about a
> man injecting male hormones into his body. . . I'm not saying
> steroids are right, but they're more right when a man's doing
> them than a woman. (Mollica 66)

Since this time steroid abuse among male athletes in general has
gained national attention and bodybuilding has recently responded
with rigorous testing and penalties for the men as well.

While the metaphorical question of beauty or the beast is embedded with the issue of steroids, concepts of femininity as a biological discourse surround the meaning of steroids. For example, Tina
Plakinger, American Champion, admits to past steroid use. She reports: ". . . when I was taking steroids I noticed that I would change
my dress style. I didn't want to wear feminine styles anymore. For
example, I hated to wear dresses. I would wear old sweats. . . My way
of walking and sitting changed too" (Vedral 220). Time does not
permit a more extensive discussion of such beliefs. Suffice it to say
that ideologies about steroids and their effects are seen through the
lens of gender and provide a narrative of biological reductionism.

Interviews with competitive women bodybuilders reveal that
the effects of increased muscularity and hardness from steroids were
desirable, while the side effects of lowering of the voice, facial hair
growth, and an overall hardness of the face were all cited as evidence
of masculinity and were negatively perceived. The culture of beauty
cannot be escaped, as notions of femininity extend far beyond relative degree of muscular development attributed to steroids. The issue of femininity and muscularity has been evident from the incep-
tion of the sport prior to the widespread use of steroids among
women bodybuilders at national and professional levels of competition.

CONSTRUCTING BEAUTY

The bodybuilder Bev Francis is an excellent case study in the
construction of beauty. As a former powerlifter, Francis displayed
phenomenal size and as one of the stars of the movie *Pumping Iron
II: The Women,* she is "the beast." She was accused of looking like a
man (Francis 97) and the judges accordingly gave her last place in
Caesar's World Cup competition in 1983. By 1987 and 1988, Bev
Francis vaulted to third place in the Ms. Olympia. As noted by one

writer, Francis learned to "throw curves instead of lumps." She has recently been described as being as "feminine as a butterfly, hardy as an ox but as delicate as perfume" (Schmidt 63).

What has Francis done to achieve this turnaround? Certainly the deliberate reduction in her muscular size and the concomitant increase in the muscularity of other competitors over time are factors. An analysis of her public persona indicates Francis additionally conformed to imperatives in the culture of beauty. Her hair has become increasingly blonde and fluffy, she wears makeup, usually competes in a pink posing bikini, and, in magazine layouts of her training techniques, appears in color coordinated spandex outfits. In addition, Francis has had plastic surgery on her nose. As noted by women bodybuilders, contest grooming is regarded as a salient variable in competition. How-to books on contest preparation point to the importance of hair, makeup, nails, and the color of the posing suit in the competition (e.g., Weider and Weider 108).

Because women's bodybuilding is for women, and women display while men act, the sport of competitive bodybuilding will continue to neutralize its masculine elements through the culture of beauty. The culture of beauty will in turn have similar consequences for women bodybuilders in terms of eating disorders and self-image problems as suggested by the data I have collected ethnographically. Interpretation from in-depth interviews with women bodybuilders suggests their self-image fluctuates with the cycles of competition and approximation to ideal contours of women's attractiveness norms. Competitors preparing for a competition actually reduce their size by dieting and losing body fat, whereas off-season body fat returns and their overall size is enhanced. This off-season image is a problem for every bodybuilder interviewed. As the women got thinner and smaller they felt increasingly better about themselves; but in the off-season their self-image plummeted as they increased body fat and wavered from the ideal.

In conclusion, while appearing contestatory and transformative, the sport of bodybuilding substantiates widespread beliefs that womanhood is enveloped in beauty written in a variety of narratives from interviews, magazine profiles, and bodybuilders' forms; and as I hope to have demonstrated, beauty inscribes concepts of femininity and even youth for the woman bodybuilder. Additionally, beauty beliefs affect women's contours through cultural norms which, in turn, are articulated with capitalism, feminism, and health reform. Finally, while the beast in women's bodybuilding is ultimately tamed by the beauty, its disruptive whispering demands attention.

NOTES

1. See McDougall (402–403) and Freund (47) for a detailed discussion of these ideas.

2. See Perper for an excellent analysis of bio-sociology and symbolism.

3. The history of beauty in the United States has been a history of an ideal generated by white elite women. Ethnic beauty as an alternative model only gained momentum in the 1960s.

4. In 1872 Darwin contended "As women have long been selected for beauty it is not surprising that some of their variation should have been transmitted exclusively to the same sex; . . . Women are everywhere conscious of the value of their own beauty; . . ." (in Freedman 181). The sociobiological argument has focused on reproductive strategies so, according to Symons, males have been differentially selected to perceive beauty (in Freedman 183). Mellon has proposed that this beauty is not generic, but that males are programed to find younger women, displaying qualities of reproductive vigor, attractive (in Zihlman 373). These views support North American age norms and ignore the cross-cultural record in which the mature woman acquires attractiveness (see Brown and Kerns *In Her Prime*) or in which male beauty may be celebrated. It is imperative to search for other factors to explain the extremes of the beauty/youth equivalency in the United States.

WORKS CITED

Annand, David. "Body or Beauty?" *Ironman* 48.8 (1989): 18.

Armstrong, David. *Political Anatomy of the Body.* Cambridge: Cambridge University Press, 1983.

Banner, Lois. *American Beauty.* Chicago: University of Chicago Press, 1983.

Blacking, John. "Preface." *The Anthropology of the Body.* Ed. John Blacking. London: Academic, 1977a. v–x.

———. "Toward an Anthropology of the Body." *The Anthropology of the Body.* Ed. John Blacking. London: Academic, 1977b. 1–28.

Bordo, Susan. "Reading the Slender Body." *Body/Politics: Women and the Discourses of Science.* New York: Routledge, 1990. 83–112.

Brain, Robert. *The Decorated Body.* London: Hutchinson & Co. Pub. Ltd., 1979.

Broverman, Inge K., Susan Raymond Vogel, Donald M. Broverman, Frank E. Clarkson, and Paul Rosenkrantz. "Sex Role Stereotypes: A Current Appraisal." *Journal of Social Issues* 28.2 (1972): 59–78.

Brown, Judith, and Virginia Kerns. *In Her Prime.* Amherst, Massachusetts: Bergin and Garvey, 1985.

Bullough, Vern and Bonnie Bullough. *Sickness, and Sanity: A History of Sexual Attitudes.* New York: New American Library, 1977.

Chernin, Kim. *The Obsession: Reflections on the Tyranny of Slenderness.* New York: Harper and Row, 1981.

Clifford, James. "Introduction: Partial Truths." *Writing Culture.* Eds. James Clifford and George E. Marcus. Berkeley, California: University of California Press, 1986. 1–26.

Devor, Holly. *Gender Blending: Confronting the Limits of Duality.* Bloomington, Indiana: Indiana University Press, 1989.

Douglas, Mary. *Natural Symbols.* New York: Vintage, 1973.

Duff, Robert W., and Lawrence K. Hong. "Self Images of Women Bodybuilders." *Sociology of Sport Journal* (1984): 374–380.

Durkheim, E. *The Elementary Forms of Religious Life.* Glencoe, Illinois: Free, 1961.

Ewen, Stuart, and Elizabeth Ewen. *Channels of Desire.* New York: McGraw Hill, 1982.

Fisher, H. Th. "The Clothes of the Naked Nuer." In *The Body Reader.* Ed. Ted Polhemus. New York: Pantheon 1978. 180–193.

Flax, Jane. "Postmodernism and Gender Relations in Feminist Theory." *Signs* 12.4 (1987): 621–643.

Foucault, Michel. *The Use of Pleasure.* New York: Vintage, 1985.

Francis, Bev. "Building My Back to Win." *Muscle and Fitness* 50.7 (1989): 97, 98, 247.

Freedman, Rita. *Beauty Bound.* Lexington, Massachusetts: Lexington, 1986.

Freund, Peter E. S. *The Civilized Body.* Philadelphia: Temple University Press, 1982.

Gaines, Charles, and George Butler. "Iron Sisters." *Psychology Today* 17 (1983): 65–69.

Gallagher, Catherine, and Thomas Laqueur. *The Making of the Modern Body.* Berkeley: University of California Press, 1987.

Guthrie, R. D. *Body Hot Spots.* New York: Van Nostrand Reinhold, 1976.

Hall, Edward T. *The Hidden Dimension.* Garden City, New York: Doubleday, 1966.

Hanks, W. F. "Text and Textuality." *Annual Reviews in Anthropology* 18 (1989): 95–127.

Hanna, Judith L. *To Dance Is Human: Theory of Non-Verbal Communication.* Chicago: University of Chicago Press, 1987.

———. *Dance, Sex and Gender: Signs of Identity, Dominance, Defiance and Desire.* Chicago: University of Chicago Press, 1988.

Haupt, Herbert A., and George D. Rovere. "Anabolic Steroids: A Review of the Literature." In *The American Journal of Sports Medicine* 12.6 (1984): 469–484.

Kessler, Suzanne J., and Wendy McKenna. *Gender: An Ethno-Methodological Approach.* New York: John Wiley & Son, 1978.

Lurie, Alison. *The Language of Clothes.* New York: Vintage, 1983.

Marcus, George E., and Michael M. V. Fischer. *Anthropology as Cultural Critique.* Chicago: University of Chicago Press, 1986.

Martin, Emily. "The Cultural Construction of Gendered Bodies: Biology and Metaphors of Production and Destruction." *Ethnos* 54.3–4 (1989): 143–180.

Mascia-Lees, Frances E., Patricia Sharpe, and Colleen Ballerino Cohen. "The Postmodern Turn in Anthropology: Cautions from a Feminist Perspective." *Signs* 15.1 (1989): 7–33.

———. "The Female Body in Postmodern Consumer Culture: A Study of Subjection and Agency." *Phoebe* 2.2 (Fall 1990): 29–50.

McDougall, Lorna. "Symbols and Somatic Structures." *The Anthropology of the Body.* Ed. John Blacking. London: Academic, 1977. 391–406.

Messing, Simon D. "The Non-Verbal Language of the Ethiopian Toga." *The Body Reader.* Ed. Ted Polhemus. New York: Pantheon, 1978. 251–257.

Mollica, Mae. "Body and Soul." *Flex* 4.2 (1986): 64–68, 71.

O'Hara, Georgina. *The Encyclopedia of Fashion.* New York: Harry N. Abrams, 1986.

Perper, Timothy. *Sex Signals: The Biology of Love.* Philadelphia: Isi, 1985.

Polhemus, Ted. *The Body Reader: Social Aspects of the Human Body.* New York: Pantheon, 1978.

Postal, Susan. "Body Image and Identity: A Comparison of Kwakiutl and Hopi." *The Body Reader.* Ed. Ted Polhemus. New York: Pantheon, 1978. 122–133.

"Preview: Ms. Olympia November 25, 1989." *Muscle and Fitness Magazine* 50.11 (1989): 74–76, 199–200.

Raskin, Donna. "The Shape of the Union." *American Health* 8.8 (1989): 31.

"The Return of the Female Form." *Cosmopolitan* (1989): 184–186.

Schmidt, Julian. "Bev Francis: The First Lady of Weight Training." *Flex* 7.5 (1989): 62–64, 87, 88.

Symons, Donald. *The Evolution of Human Sexuality.* Oxford, England: Oxford University Press, 1979.

Tanny, Mamsy. "Cory Makes 5." *Muscle and Fitness* 50.3 (1989): 88–97.

Tress, Daryl McGowan. "Comment on Flax's Postmodernism and Gender Relations in Feminist Theory." *Signs* 14.1 (1988): 196–203.

Turner, Bryan S. *The Body and Society.* Oxford, England: Basil Blackwell, 1984.

Vedral, Joyce. "Ms. O Contestants: Sexier This Year?" *Muscle and Fitness* 50.7 (1989): 124–127.

Weider, Betty, and Joe Weider. *The Weider Book of Bodybuilding for Women.* Chicago: Contemporary, 1981.

6

As You Stand, So You Feel and Are: The Crying Body and the Nineteenth-Century Text

Why is it that we resent being made to cry? To be more specific: why do we (I mean educated, self-conscious, theoretically sophisticated, socially, and culturally progressive readers and viewers) tend to respond to "sentimental" texts with dismissive annoyance, even (or especially) when they have the power to "get to us"?

My "we" speaks from inside academia, rather than for American culture at large. And, to be sure, the mainstream academic community and the mass media certainly differ over sentimentality in at least one important respect: writers of journal articles and university-press books seldom if ever produce texts calculated to make audiences cry, while those who contribute to the mass media rely on what they might call "tear jerkers" as a continually lucrative product. But the official pronouncements and unstated assumptions about sentimentality are surprisingly consistent from the media to the academy. The media tell us, for instance, that when the audience of a high-art theatrical production weeps, the tears are the sign of a tragedy's power and validity: *Newsweek* says of a recent Joseph Papp production of *The Winter's Tale* that the actors "seem to be as overwhelmed as the audience, and tears are plentiful onstage and off."[1] Had this remark been made about *Steel Magnolias* or *Dad*, its implications would almost certainly have been negative. But we are talking about Shakespeare. Therefore, "this frank, immediate emotional reality is one of the strengths" of the production, according to this arbiter of middle-class taste. If a popular text (especially a film) makes actors and audiences cry, by contrast, the press's typical reaction is to assign mechanistic, derogatory labels to it and to generalize about its "manipulative" characteristics.

As an example, consider a recent article for Gannett News Service by Julie Hinds called "Get out your Handkerchiefs" (the title alludes to the Bertrand Blier movie that parodies sentimental domestic dramas). Noting the recent surge of sentimental Hollywood films, Hinds offhandedly demonstrates every cliché prejudice against the genre imaginable. All her quotations of authorities come from commentators who call such films tear jerkers and "weepers"; her own analysis refers to "getting all choked up" and "tugging . . . on . . . heartstrings." She explains that "women's weepies" have traditionally been "marketed at females, who received cathartic enjoyment out of watching someone else overcome tribulations." Never mind that the essentializing term "females" places her own resisting voice outside the realm of her sex. To provide closure for her article, she borrows a man's voice anyway, quoting a male critic who says of the sentimental effect: "That's really stupid. It doesn't ask the audience to do anything except get run over by a tractor." Hinds adds, "And bring a lot of Kleenex." Hinds's tone is light, her vocabulary slangy; her article is supposed to be amusing. But if the spectacle of the woman crying at *Terms of Endearment* or *Camille* (two of Hinds's examples) is funny, why shouldn't the audience at the Shakespeare play be open to similar ridicule?

Of course, feminist academics have been tackling the problem of the critical double standard now for nearly twenty years. Within the academy, such critics have gone some distance in the past decade toward "rehabilitating" sentimentality in certain contexts, particularly in nineteenth-century American women's literature.[2] Traces of the distaste for being made to cry still surface, however, in what even the most progressive critics say—or don't say—about sentimentality.

Beginning with the premise that a prejudice against sentimentality—and particularly against textually induced tears—pervades American cultural criticism inside the academy as well as outside, this essay seeks to locate the roots of that prejudice in a powerful but seldom-challenged model of the relationships among emotions, the body, and texts. Certainly, the dismissal of sentimentality, not to mention the mode's perverse habit of persisting in American culture despite that dismissal, can be traced to many influences having to do with the intertwining histories of gender, class, and aesthetics. But to find a starting place, I will argue that one reason we dislike senti-mental texts which make us cry is that we cling to an idea of the body as a repository of "real" emotions, a reservoir of passions that high art (such as classical or Shakespearian tragedy) can legitimately tap into and that sentimental texts can divert for exploitative or commercial purposes, thus rendering the emotions somehow less "authentic."

Among the implications of this "reservoir" model of the body and its relation to texts is the idea that the experience of elite au-diences (like the crowd at the Papp production) is more genuine, more "frank [and] immediate" than the emotional experience of the popular audience having tears wrung out of its eyes by such "exploi-tative tearjerker[s]" (as one review has it) as *Stella*[3] or—to bring up titles of classics in their own genres—*Terms of Endearment* or *Dark Victory*. I want to challenge this class-based (and, I will argue, gen-der-based) differentiation between the physical experiences of elite audiences and popular ones.

In so doing, I do not intend to make any essential distinction between the audiences of stage plays and the audiences of films: in his own time Shakespeare's productions served a popular function they no longer have on stage, and filmed versions of his plays have not necessarily been either given over to "sentimentality" or devoid of it. The distinction I am making between elite and popular au-diences is located in the present, and in making it I assume that the elite audience has more flexibility than the popular one: whereas my colleagues and I might as easily go see *Steel Magnolias* as attend a Papp Shakespeare in the Park show, the woman who cuts my hair (who tells me she cried often during that particular movie) has no interest in attending even an inexpensive Shakespeare play, or, for that matter, in going to see Kenneth Branagh's recent film of *Henry V*. From inside the academy, it is easy enough for us to see how my haircutter's interests have been socially constructed, and if we want to take a conventional stance on her tastes, we can assume with Hinds and her experts that this woman's tears serve only to alienate and disempower her, to keep her down. What is less obvious from where we stand, however, is that our own academically sanctioned response and reaction to sentimentality is also constructed, that it arises not from some "natural" aversion to certain textual strategies, but from a historically specific prejudice against sentimentalism per se. I will argue here that this prejudice has its roots in our culture's assumptions about what the "nature" of the body, and its relations to emotions, is supposed to be.

In the following sections, then, I will elaborate upon four points toward supporting this argument: (1) that the longstanding critical aversion for sentimentality lives on, even among progressive critics of American fiction; (2) that common assumptions about art, emotion, and the body drawn from Freud and Aristotle are metaphors and theories, not essential truths; that they can be placed in their respective historical moments and that they have the potential to be replaced by alternative models of the relations between art and emotion; (3) that nineteenth-century ideas about the psychophysiology of stage acting provide an example of such an alternative model, one which can illuminate that period's attitudes toward sentimental fiction; and (4) that our inherited prejudice against being made to cry—and literary criticism's evident reluctance to dismantle that prejudice—can be attributed to the fact that crying is coded as "feminine" in American culture, as it has been since at least the last quarter of the nineteenth century.[4] To come to a new appreciation for textually induced crying, then, would mean realigning attitudes about gender, as well as attitudes about class, in criticism of nineteenth-century American literature.

I.

My first authority for claiming that "we dislike being made to cry" is of course my own experience, modified by the anecdotal testimony of colleagues and students with whom I have discussed this matter. To date, no one has done a systematic study of readers' attitudes toward textually induced tears, a study that could go in its documentary persuasiveness beyond my personal report and your individual inclination to agree or to refuse to identify with how I say I feel about crying over texts.[5] I am relying, then, mainly on myself when I talk about "the reader," but a few, more general studies of readers' experiences—such as Helen Taylor's survey of admirers of *Gone with the Wind* and Barbara Sicherman's analysis of the reading habits of a Victorian American family—do provide some useful clues about other female readers' reactions to textually induced tears.

For me, the relative familiarity or newness of a text seems to have no bearing on my response if the text happens to be one of those that "get to me." Like the women Taylor interviewed, whose copies of their favorite novel are "now well thumbed [and] tear-stained" (35) and who gather ritually to cry together over repeated video viewings of the film, I can expect to be made to cry, quite consistently, by certain texts. Some passages of *Uncle Tom's Cabin* invariably bring tears to my eyes, as do some scenes near the end of *King Lear*. I am embarrassed, though—humiliated even—to admit (in a context as formal as this one) that two texts so different in reputation have so similar an effect on me.

And am I not typical, at least among contemporary feminist academics? I pride myself on my intellectual toughness, my ability to take apart any text and propose explanations for how and to what ends it was put together. Yet I am perfectly capable of sitting before a screen—watching perhaps *Terms of Endearment* (which I loathe), perhaps *It's a Wonderful Life* (one of my favorites)—or even watching a commercial for Kodak or AT&T, and experiencing the same physical response I feel when Lear and Cordelia are reconciled. Even the most crassly sentimental text can make me cry. And I resent it— as I suppose you do, too—if you are willing to acknowledge similar susceptibilities. For readers who have been trained under the modernist aesthetic of defamiliarization and alienation effects, there is something objectionable about obviously commercial texts that so readily and mechanically arouse emotion: it's too easy; it must not be "authentic." But the postmodern perspective, which allows for skepticism about traditional criticism's assumptions that some feelings are "genuine" or authentic and others are not, puts us in the position of being able to ask: Is not all interior experience to some degree socially or culturally constructed? Why, then, should we perpetuate this modern prejudice against sentimental effects in art?

For some critics, as for the average movie reviewer, the prejudice is a natural one, which passes without examination. You can pick up any non-feminist reference work on nineteenth-century American fiction almost at random to find evidence of this. Witness Herbert F. Smith's *The Popular American Novel: 1865–1920* (1980), which, for example, only briefly mentions that famous inspirer of feminine tears, *Little Women* (1868) to acknowledge the popular success of this novel "and the various sequels it spawned." Notice how the repellant metaphor removes the agency behind the novel's reproduction from the woman writer, attributing it to the text itself, which is supposed to have dropped its sequels into the world by some biological process, as a fish drops eggs. Louisa May Alcott does get grudging credit from Smith for "having dealt with middle-class life," but that gets canceled out by Smith's next (and final) pronouncement on the sentimental domestic novelist: "her use of realism to increase sentimentality was too perverse to be favored by any but children and gushing girls" (47). The formulation is odd, since realism in other literary contexts is a value for Smith. But he explains early in his book that sentimentality is a "disease," and that the "discriminating critic" must act as a "pathologist" to trace its symptoms and—presumably—to cure it (19), even where it coexists with realism.

The dismissive sexism of Smith's position on nineteenth-century women writers and their readers is so blatant as to be, in itself, uninteresting. His remarks are useful here, though, in at least three ways. First, his metaphor of infection transmutes the text into a body, which has the effect of deflecting attention from the literal body of the person (the woman, he assumes) who holds the book and reads. Second, this thinking about texts in terms of bodies, is, as I will explain below, a negative variation on nineteenth-century American critics' ideas about the way texts and sentiments operate. And third, Smith's negative assumptions—perhaps predictable in their context—also operate more subtly and more surprisingly in criticism which presents analyses of sentimental texts in less openly prescriptive terms.

According to Nina Baym's comprehensive analysis of nineteenth-century critics' criteria for judging fiction, "all good novels were expected to contain pathos," a term "used by reviewers to denote an emotional response . . . characterized by pity, sadness, or sorrow—'to dissolve the heart in tears'" (*NRR* 140). Baym emphasizes that the reviewers she studies took an "affective" approach to the novel (149), and that their attention was focused not on a work's thematic content or specific techniques, but rather its emotional impact upon individual readers (142). In their view, the best novels elicited laughter and tears. Baym locates a pattern of medicinal metaphors among individual reviews, and attributes them to "a conception of the novel as a substance taken into the body, there to work an effect beyond the reader's control" (58). For the nineteenth-century reviewer, then, sentimental passages in a text were part of the text's machinery for "working an effect" in a reader's body, and if that effect were one of "pathos," the result would be to the benefit of the reader. Consider the contrast with Smith's staunchly modern metaphor, which overlooks the reader's body (literally present though it is in any act of reading) and transforms the text into a metaphorical body instead. In Smith's model, sentimentality is the disease infecting the body of the text, whereas in the nineteenth-century model, textual pathos was the medicine for the body of the reader. For Smith, the heroic medical efforts of the critic/pathologist who expunges the disease take center stage; the crying body of the girl reading Alcott gets erased, with a sneer, from the picture.

Certainly Baym is very careful to acknowledge the historical contingencies at work in the evaluation of fiction, and though she wryly declines to "spend much time on the issue of how it might be that people enjoyed being made to feel bad" (141) she scrupulously resists the kind of dismissal of nineteenth-century values that Smith practices. In an earlier book she cautions against using sentimental as "a term of judgment rather than of description," because such judgments, usually negative, "are culture-bound": the critic of a later time who uses the term disparagingly is refusing "to assent to the work's conventions" (*Women's Fiction* 24). But even Baym is not immune from her era's negative assumptions: why does she see her subjects as enjoying "being made to feel bad"? No excerpt that she cites from her nineteenth-century reviewers indicates that her subjects found anything "bad" in the feelings they are writing about. In other words, to avoid culture-bound dismissal of the popular nineteenth-century novel, we need to assent to conventions of the reading process, as well as of characterization, story, or style.

Far less willing than Baym to accept the conventions of her subject matter is Janet Todd, who has chronicled the philosophic and aesthetic history of British sentimentalism in *Sensibility: An Introduction* (1986). Todd's attention to the productions of the cult of sensibility yields a thorough and helpful survey of that eighteenth-century mode of writing. But she takes it for granted that twentieth-century readers and viewers, with their "taste . . . for the ironic and self-reflexive in literature" must necessarily find the mode "repellant" (142). She means, of course, *elite* twentieth-century readers and viewers. Skipping ahead to Victorian America, she concludes her book with the famous passage from *Uncle Tom's Cabin* where Eva dies what Todd chooses to call a "public and redemptive but politically impotent" death (149). As I will argue more fully below, seeing Eva's death as "politically impotent" requires limiting one's view to what happens inside the fiction, rather than

attending to the fiction's impact upon actual readers and their potential involvement in political change. But for now, let us focus on
Todd's negative attitude toward the sentimental response per se.

Todd concludes her book with what appears to be a set of
alternatives for how readers might respond to sentimentality. Close
analysis suggests, however, that Todd's alternatives boil down to
something like this: "You can hate sentimentality, or you can hate
it." Labelling the style of Eva's deathbed scene "overcharged," Todd
remarks:

> Depending on period and personality, readers of such prose can
> join Godwin and Coleridge in lamenting the prestige of feeling
> in culture, or they can with Doris Lessing's character in *The
> Sentimental Agents of the Volyen Empire* (1983) see such flam-
> boyant sensibility as a destabilizing addiction and a glamorous
> corruption; they can turn away in embarrassment or they can
> choke and cry. (149)

The parallel structure of this sentence makes for an elegant and
resonant ending to Todd's book, but its logic is not clear. "Readers . . . can . . . [1] lament . . . or . . . [2] see . . . addiction and . . .
corruption; they can [1] turn away . . . or . . . [2?] choke and cry."
The syntax suggests that sentimentality can [1] embarrass us, as it
did the male Romantics, or it can [2] appear to us in the light of
addiction and corruption, as it does in Lessing's anti-sentimental
novel. Hanging on at the end of the sentence is the second [2] above,
which does not parallel the first [2] and has no place in the sentence's logic: "they can choke and cry." Is that what readers do, who
see sentimentality as addiction and corruption? Is there no alternative for the *other* reader who finds something positive, or possibly
even empowering, in the sentimental response? Significantly, Todd
draws upon the same set of disease metaphors that Smith uses to
talk about textual sentimentality, though this time it's the reader's
body that is subject to addiction and corruption. Though Todd acknowledges response must depend "on period and personality," she
nevertheless leaves no room in her period for the personality who
might like sentimentality.

Todd's aversion aligns her with commentators who see sentimentality as politically impotent, an attitude she attributes in particular to Mary Wollstonecraft (132). The idea that someone who
cries over representations of injustice, poverty, and pain will be necessarily helpless to do anything about their "real world" referents is

pervasive in modern criticism, as we have seen in the case of the move reviewer who asserts the sentimental text "doesn't ask the audience to do anything." But the idea is peculiarly unsuited to the experience of nineteenth-century middle-class Americans, especially women. Alice Hamilton (whose family's reading is the subject of Barbara Sicherman's recent research) reports that her father considered her Victorian American mother to be subject to "sentimentality" which "he hated"; her mother's favorite literature contained decidedly "pathetic" material (including Gray's *Elegy* and George Eliot's *The Mill on the Floss*). But if Mrs. Hamilton could cry over a book, she could also take responsibility for action. The daughter repeats the mother's motto in the face of social injustice: "There are two kinds of people, the ones who say 'somebody ought to do something about it, but why should it be I?' and those who say 'somebody ought to do something about it, then why not I?'" (32); Alice Hamilton places her mother in the second camp. For a woman holding such attitudes, the emotional affect aroused by a book was its means of making its subject matter personal to her: the sentimental response was a sign that the fictionally represented experience was registering in her body. Evidently the act of crying was not, for her, a vent for passions that might more productively have been spent elsewhere; instead, it fanned the flame of what her daughter admiringly calls her "enthusiasms" and "indignations." Mrs. Hamilton's behavior, then, is radically opposed to the assumption Todd shares with the movie reviewer, who, as we have seen, claims females "received cathartic enjoyment out of watching someone else overcome tribulations."

Of course, the most progressive critics acknowledge that sentimentality had political and cultural work to do in nineteenth-century America. Jane Tompkins has argued eloquently about *Uncle Tom*, among other sentimental novels, that it "represents a monumental effort to reorganize culture from a woman's point of view" (124), and that while it obviously never fully achieved that goal, it did have an enormous cultural and political impact on its large, enthusiastically responsive Northern audience. Tompkins attends primarily, though, to the text's representations of weeping—and to the assumptions about religious "truths" its audience presumably held—without really talking about the audience's own propensity for crying. Certainly, one can more readily come up with data for a study of how crying is depicted in texts than for an inquiry into how crying gets inspired by them. Despite the ontological difficulties presented by the question, however, I think it is time to extend the study of nineteenth-century tears to the extratextual realm. I will show, in Part III below, that representations of weeping are not necessarily the only, or the most potentially effective, means that sentimental texts use to inspire readers' tears. I contend, too, that to focus on weeping in the text, rather than acknowledging the crying reader in one's analysis, is to evade the referent of our culture's discomfort with this phenomenon: the flesh-and-blood body of the crying woman (or "gushing girl") who reads. Jane Tompkins is responsible for disseminating the idea that modern criticism categorically dismisses texts that "do something" to readers—she has long been a leading champion in the fight against such previously unexamined modernist critical prejudices. But even the unembarrassable Tompkins (it took a lot of intestinal fortitude to write "Me and My Shadow," which exposes the critic's own emotional experience of academic life in an unprecedentedly bold way) turns aside from the spectacle of crying readers.[6]

A project following close on the heels of *Sensational Design* in its efforts to see popular nineteenth-century novels in a rehabilitative light, Philip Fisher's *Hard Facts* looks more deliberately than Tompkins' book at the physical, emotional effects sentimental texts can have on their willing readers. For Fisher, the act of crying is part of the reader's participation in the novels' project of "familiarizing" formerly marginal and oppressed groups, such as slaves, children, or the insane. Following a model proposed by Jean-Jacques Rousseau, Fisher reduces several of the tear-inspiring scenes in *Uncle Tom's Cabin* to a basic paradigm: an immobilized witness watches a monster violently separate a victim from the victim's loved one; unable to intervene, the witness has no recourse but to weep. *Uncle Tom* is full of reiterations of this pattern, different from Rousseau's model only in that the characters' tears well up during narrations of the sad events as well as during their direct enactment. (I am thinking, for instance, of the scene where Senator Bird's whole household weeps while Eliza tells them her story, or the scene where young George Shelby is overcome with tears when he tries to write to his mother about Tom's death.) Fisher suggests that the reader of these scenes is in a position analogous to Rousseau's witness: unable to intervene in the oppression, to bridge the gap between text and world, a reader who weeps is, according to Fisher, venting helpless frustration.

Taking sentimentalism on its own terms, Fisher's theory positions itself as a defense of the mode within its historical and cultural context. As he puts it, "by limiting the goal of art to the revision of images [of oppressed persons or groups] rather than to the incitement to action, sentimentality assumes a healthy and modest account of the limited and interior consequences of art" (122). Embedded in this conclusion, however, are at least two strong signs of the kind of negativity towards sentimentalism that is so much more clearly in evidence in other critics' work. First, Fisher's model may account for the "tears of defeat" associated with sentimentality, but it does not address the "tears of triumph" (James Smith 15, 56), the "good cry" that dominated sentimental reading experiences later in the nineteenth century and that contribute, for that matter, a large part of the emotional impact of *Uncle Tom* itself. Second, Fisher's theory supports the idea that crying over fiction equals political impotence. His argument forces him to relegate to a footnote the truism that *Uncle Tom's Cabin*, in his words, "is perhaps the single most effective political work of art in the history of literature"; his note can only shrug this off as a "paradox" (182). In other words, Fisher's model works beautifully in theory, but faced with the historical fact that crying readers did take action after having read the novel, the theory falters. I suggest the problem is the critic's reluctance to look too directly at the crying audience, and his model's reliance on the notion that textually induced tears are necessarily cathartic. Perhaps the theory of catharsis, influential as it has been in criticism since Aristotle's formulation of the idea, is not the most appropriate working model for explaining the experiences of nineteenth-century readers.

II.

Aristotle's *Poetics* takes for granted certain theories about how emotions operate in the body, theories which persist—as we have seen in Todd and Fisher—in literary criticism as they do in traditional Freudian psychoanalysis. Quite casually, Aristotle explains in Part VI of the *Poetics* that one of tragedy's defining features is that it will "through pity and fear [effect] the proper purgation of these emotions" (27). This theory of "catharsis" has been taken in the twentieth century to mean that audience members' bodies contain a given quantity of pity and of fear, and that the experience of weeping at a tragedy constitutes the "proper purgation," the healthy venting or draining of those emotions. According to Aristotle's more detailed description of the effect in Part XIII, "pity is aroused by unmerited misfortunes, fear by the misfortune of a man like ourselves"

(33). In other words, the "fear" effect relies on the audience's identification with the tragic hero and the "pity" effect depends on the ethical and moral conclusions the audience draws upon the hero's fate. Both these effects are, then, strictly culture-bound: they cannot be assumed to work upon audiences who—for whatever reasons having to do with ethnicity, nationality, class, gender, or era—do not identify with the hero or who do not see his fate as unmerited. In any given tragedy, then, the effects themselves are not essential or eternal, but wholly dependent upon the psychological and moral orientation of the audience. I want to suggest taking this relativist revision of Aristotle's model a step further, to ask: are its assumptions about the reading process necessarily more universally transportable from text to text or era to era than the effects it seeks to describe?

Evidently Freud would have answered "yes," as he, too, seems to have taken the catharsis model for granted. In their explanation of the mechanics of hysteria, he and Breuer posit that a traumatic memory can only healthily fade if "there has been an energetic reaction to the event that provokes an affect," reactions being "the whole class of voluntary and involuntary reflexes—from tears to acts of revenge—in which, as experience shows us, the affects are discharged" (8). The reaction to the trauma is "cathartic" if it is "adequate," but "if the reaction is suppressed, the affect remains attached to the memory" (8). If, however, "there is no such reaction, whether in deeds or words, or in the mildest cases in tears, any recollection of the event retains its affective tone" (8). In other words, unless the emotion is purged or discharged, it remains dammed up in the body, and until the proper catharsis occurs, it will continue—through hysterical manifestations—to affect the body's functions. As Carol Tavris summarizes this view, "Freud imagined that the libido was a finite amount of energy that powers our internal battles. If the energy is blocked here, it must find release there" (37).

As Tavris is quick to point out, however, Freud's language of energy and catharsis is metaphoric, and having never been "scientifically proven," remains a way of talking about how emotions function, rather than a definitive account of that process. Critiquing psychotherapeutic practices which take literally the idea that "emotional energy is a fixed quantity that can be dammed up or, conversely, that can 'flood' the system" (22), Tavris calls this "ventilationist" view into question. According to recent work on the expression of hostile feelings, the "reaction" through "affect" associated with angry feelings may not function cathartically in all cases, but may sometimes only rehearse the affect. Since, as Tavris points out, "any emotional arousal will eventually simmer down, if you just wait long enough" (122), people who vent hostile feelings will eventually feel better, as will people who do not vent them. But the people who act upon their feelings will "be acquiring a cathartic habit" (126). To shout or hit someone, she explains, is not necessarily a vent for anger—it can instead be seen as a reinforcement of shouting or hitting behavior. If this is true, then perhaps to cry is not necessarily a purgation of fear and pity.

My point is not to try to say whether Freud's model or those that challenge it are "right." I am interested instead in the fact that such models are "historicizable," that they have their respective places in history and, doubtless, their influences upon discourse about emotions and art. This observation opens the possibility that other accounts of the body's relation to emotion could alter our view of the role texts might play in stirring up feelings, or more precisely, might put us in the position of grasping how another era's notions on the matter might have colored its literary tastes. Aristotle and the twentieth-century critics who do not question his assumptions say that tragedy is cathartic and therefore good. I want to suggest that sentimentality, despite what Todd and Fisher imply, is pointedly *not* cathartic, but that it rather encourages readers to rehearse and reinforce the emotions it evokes. Does that make sentimentality bad? Within an Aristotelian framework, perhaps; but not from the perspective of Freud's predecessors, whose notions of psychophysiology differ markedly from the model I have been describing.

III.

The twentieth-century idea that the emotions somehow inhabit the reservoir of the body requires conceptualizing a binary split between the body and the mind. Joseph Roach, whose work on the history of the theater raises this issue, points out that this mind/body split was not current in the nineteenth-century psychology and physiology. The dominant idea of the dynamics of mind and body during the Victorian period in Europe and America was a version of monism, elaborated by George Henry Lewes in his capacity as naturalist (and applied in his capacity as drama critic).[7] Monism sees the mind and body not as a duality, but a continuum. Lewes organized that continuum into four levels. First is the level of "neural tremors," the sensory data received by the body; second, the level of sensation (made up of groupings of tremors); third, the level of image (made up of groupings of sensations); and fourth, the level of symbol or idea. The four levels can act interchangeably as stimuli, one upon another: for example, the idea can evoke the sensation, or the neural tremor can evoke the image. As Roach explains, Lewes's model leads inexorably to William James's conviction that "the physiological manifestation of the emotion *is* the emotion . . . it is because we weep that we feel sad, it is because we . . . tremble that we feel afraid" (192).

As an analogy to the different Victorian and modern views of the reading process, let us consider the difference between the two periods' dominant ideas about acting. "According to the principles laid down by both Francois Delsarte and William James," reports Martha Banta, "physical acts initiate emotions and make them authentic. When your body assumes a pose in a particular setting, that pose becomes your sign. As you stand, so you feel and are" (647).[8] The Delsarte to whom Banta refers was a French philosopher (1811–1871) whose disciples codified his theories of pose and gesture, adapting them into conventions to be used by melodramatic actors and artists' models. Those conventions were entirely opposed to popular modern notions about how actors exploit the relation between the body and the emotions. A twentieth-century "method" actor, following the precepts of Stanislavski, attempts to "get into character" before performing a part. This entails the actor's concentrating on "becoming" the character internally, in order to represent him or her bodily. In the practice of such method actors as Marlon Brando, Dustin Hoffman, or Robert DeNiro, becoming a character often means drawing upon memory of lived experience to recreate emotions the actor has felt in circumstances comparable to those being represented in the diegesis. In other words, the method actor draws upon that reservoir of stored emotion to vent it during the performance. In this modern style of acting, the gesture—be it a subtly raised eyebrow or a dramatically brandished arm—serves to express the feeling which the actor who is "in character" is "really" experiencing.

Delsarte's idea of acting stands this concept on its head: Instead of seeing gestures and poses as expressions of real interior states, he saw them as vehicles for producing those states. In Delsarte's scheme, each body part (the hand, the torso, the forearm, etc.) is connected to *L'Esprit, L'Ame,* or *La Vie,* categories his followers translated as "the Mental, the Moral, and the Vital." The mental zones of the body are linked to the operations of the intellect, the moral zones are linked to emotions, and the vital zones are linked to sensation. The chest, for example, is a mental zone (the Delsarteans do not explain why), and the sob (a kind of breathing that puts the chest into a violent or "hysteric" motion) is—according to one of Delsarte's adapters—a bodily motion that actors must use "very carefully," because it both signifies and produces "a mind unbalanced" (Stebbins 288). For Delsarte, then, the gesture not only expresses the emotion and the state of mind, it brings them into being. Hence the actress positioned with face upturned, right hand raised to the forehead, left hand clenched and arm extended behind her, eyes closed and brows in an inverted 'v,' will feel the sentiment her pose represents in a Delsartean handbook: "Mine woes afflict this spirit sore" (Banta 645). The feeling will be "realized"; the pose will make the actress feel it, and her bringing it into physical presence will transmit its effect to the spectators.

Delsarte's formulation, of course, precisely reverses the familiar reservoir model on which the principle of catharsis is based. The modern view says we weep because we feel sad, and in weeping we get that sadness "out of our systems." Based on essentialist assumptions about "human nature," the modern view leads to the evaluation of some feelings as more genuine, more fully human than others—which leads inexorably to the denigration of the emotional experience of persons whose cultural or social marginality marks their feelings as "different." Perhaps the most astonishing aspect of the Victorian view is the easy alliance it forms with postmodernism, in that both reject essentialism as a way of accounting for interior experience. Furthermore, the Victorian idea that we feel sad *because* we weep puts a radically different spin on weeping's ultimate effect. This means that from the Victorian perspective, crying over *Uncle Tom's Cabin* or *Little Women* did not drain a reservoir of stored feelings, nor did it debilitate readers from taking action in the extratextual world. Instead, crying created and promoted the feelings, which then might presumably serve as goads to action.

If nineteenth-century sentimental novelists were not consciously aware of such theories of the relations among bodily action and emotion, they seem at least to have been operating under assumptions similar to those that inspire the theories. The sentimental novelist, like the Delsartean actor, relies on a set of "mechanical" exercises (as Delsarte's detractors called his gymnastic drill for actors) to get the body of the reader into the pose that would generate real compassion, the pose of weeping. The narrator of the nineteenth-century sentimental novel puts the reader through the drill techniques established in such eighteenth-century precursors as Susanna Rowson's *Charlotte Temple* (1791). Although a critic as historically minded as Cathy Davidson suggests that Rowson's novel worked on its audiences' emotions through their identification with its heroine (170), I think that identification (another post-Freudian model we frequently impose on the reading experience of earlier generations) may not have been the only process at work in Rowson's readers.[9] Many of the "mechanical" techniques still operating in today's tear jerkers must have been at least partly responsible for that first American bestseller's emotional impact.

IV.

These "mechanical techniques" in fiction are so common as to be easily recognizable. Anyone who has sat in a crying audience at a film knows that the response to particular moments in the diegesis is not usually idiosyncratic: the person who cries at a film is likely to find, upon looking around, that others are crying, too. As my colleague Roxanne Lin pointed out to me, the textual strategies that inspire tears are a set of cues that are recognized within a given culture: "Like when someone opens a door—you know to step through it." Unlike the opened door, these cues are not equally effective for every receiver: many readers resist the sentimental effect, some so effectively as not to be aware that the cues exist. In pointing out the cues that operate in nineteenth-century sentimental texts (many of which are effective for the 1990s audience accustomed to their variations in melodramatic film and soap opera), I am relying on the consensus of "susceptible readers," students and colleagues who recognize the experience I am pointing to.

To isolate and describe those cues that nineteenth-century texts use to try to evoke tears, I rely on the vocabulary of the structuralist narratologists who distinguish "story" (*what* happens in a diegesis) from "discourse" (*how* the story gets rendered). I follow the example of Gérard Genette, among other narratologists, in focusing on the conventions of discourse (e.g., narrative voice, perspective and focalization, and repeated plot patterns) rather than interpreting the details of the story itself. Using narratology presents a problem, though, for the theorist with an interest in novels' effects upon readers, because structuralism traditionally views the text as discontinuous with the extratextual world, completely cut off from it.

As I have explained above (and in other contexts), I agree with the many theorists who argue that criticism can no longer afford to overlook the physically present body of the reader who brings the text into being through the act of reading, for to overlook the actual reader is to elide the crucial differences that extratextual factors such as class, race, and gender can make in the reception and evaluation of fictions.[10] In describing seven patterns I discern in the narrative discourse of sentimental novels, then, I will be orienting each toward its habitual effect upon the susceptible and cooperative reader, in this case the middle-class Victorian American woman—or any contemporary reader willing to try to emulate her reading experience. Those seven patterns follow:

1. Such novels as *Charlotte Temple, Uncle Tom,* and *Little Women* employ narrative voices that draw upon the connotative power of poetic devices to heighten the emotional impact of the prose. To be sure, there is nothing intrinsically emotional about poetic language. But these devices would have been familiar to the nineteenth-century woman through the lyrics of hymns, the poems in keepsake books and literary collections, and the commemorative poetry family members wrote to console one another for bereavements. In each of these cases, poetic language would carry heavily emotional connotations through its association with spirituality, friendships, family relationships, and the loss of loved ones.

The most common poetic devices in sentimental novels are the most purely mechanical. They include, for instance, alliteration: "A sudden sickness seized her; she grew cold and giddy, and putting it into her husband's hand, she cried . . . " (*CT* 90); "But a bird sang blithely on a budding bough, close by, the snowdrops blossomed freshly at the window, and the spring sunshine streamed in like a benediction over the placid face upon the pillow—a face so full of painless peace that those who loved it best smiled through their tears . . . " (*LW* 392). Equally mechanical is the reliance on punctuation that emphasizes emotional affect in the narrator's utterances (including exclamation points, dashes, and italics).

Less obviously, sentimental prose plays with tropes of presence and absence to subvert the logical relations between the imagined world of the diegesis and the physical world inhabited by the reader, blurring the lines between them for emotional effect. For instance, sentimental narration can use apostrophes to bring the characters into the same plane of reality as that inhabited by the readers, whom the narrator also frequently addresses: ("Even so, beloved Eva! Fair star of thy dwelling! Thou are [sic] passing away. . ." *UTC* 383). This self-consciously poetic style is, of course, what Todd means when she calls the language of Eva's deathbed scene in *Uncle Tom* "overcharged."

2. The embellished style runs parallel to a theme common in sentimental novels (and their close relative, Gothic novels) suggesting language, in the form of mere words, is inadequate to convey the depths of emotion the characters and narrator are presented as feeling. This theme is transmitted in the narrative discourse, both through reiterated scenes of characters' being pushed beyond speech by their feelings, and through narrators' remarks. In *Charlotte Temple*, for instance, the characters are frequently too overwhelmed by emotion to be able to utter words: "his heart was too full to permit him to speak" (24); even more often, their bodies are overpowered by their emotions, and they "fall prostrate [or senseless] on the floor" (16, 48, 85, 109). The sentimental narrator reinforces this theme by frequently claiming that the emotions she must report cannot be rendered in language, but must be "unnarrated," as Gerald Prince calls the technique of naming that which cannot be told in a story: "[she] gave vent to an agony of grief which it is impossible to describe" (*CT* 86); "I don't think I have any words in which to tell the meeting of the mother and daughters; such hours are beautiful to live, but very hard to describe, so I will leave it to the imagination of my readers" (*LW* 187). Such unnarrated passages are a signal to the reader to fill in the blank with the emotions for which the narrator cannot find words. The cooperative reader must follow the cues and take an active part in the co-creation of the scene's emotional power.

3. Sentimental narrative discourse requires a particular handling of "internal focalization," Genette's term for the technique of limiting narrative perspective temporarily to one character's consciousness, despite the fact that the character does not speak the narrative. Scenes in sentimental novels tend to be focalized either through victims (such as the many slaves separated from their families in *Uncle Tom*) or triumphant figures who have formerly been represented as oppressed (such as Beth March in the joyful moment of her thanking Mr. Laurence for her piano.) This focalization invites the reader to participate emotionally from the subject-position of the oppressed. Sentimental novels can use embedded first-person narratives to achieve this effect, as when Cassie narrates to Uncle Tom the miserable story of her life as a "quadroon" slave (*UTC* 514–522). More often, the "omniscient" narrative focus simply shifts to the perspective of the sufferer, rendering the scene as he or she sees it, for instance, in the chapter of *Little Women* called "Amy's Valley of Humiliation," in which the youngest of the heroines is physically

punished at school. As Fisher has pointed out, the focalization sometimes comes through intermediary figures who are not, themselves, directly oppressed—such as Eva in *Uncle Tom*—but it is seldom if ever granted to those who oppress the protagonists in the fictional world. This careful limiting of the narrative point of view to those who suffer and triumph after tribulation can effect a powerful pull on the emotions of a susceptible reader.

4. The narrators of sentimental novels frequently use earnest, direct address to a narratee, calling upon him or her to recognize parallels between lived experience and the situations represented in the fiction. This, of course, is the notorious "preachy" tendency of sentimental novels that so many twentieth-century readers cannot abide. Direct address to the reader is, however, central to the sentimental novelists' project, in that it does not leave the matter of identification up to the details of the story alone. The sentimental narrator (usually an example of what I have called elsewhere the "engaging narrator") enforces comparisons between a reader's experience and characters' experiences which readers might not "naturally" have made themselves.

Rowson does it often: "My dear young readers, I would have you read this scene with attention, and reflect that you may yourselves one day be mothers. . . . Then once more read over the sorrows of poor Mrs. Temple, and remember, the mother whom you so dearly love and venerate will feel the same. . ." (*CT* 54). Stowe does it even more often: "In such a case, you write to your wife, and send messages to your children; but Tom could not write" (*UTC* 228); "And if you should ever be under the necessity, sir, of selecting, out of two hundred men, one who was to become your absolute owner and disposer, you would, perhaps, realize, just as Tom did, how few there were that you would feel at all comfortable in being made over to" (476). Any given reader's ability to answer to such an appeal would depend on that reader's willingness to be constructed by the narrator's utterance (does Rowson's reader have to be female, to feel the thrust of her address? Does Stowe's have to be male?). But those readers who could respond, whether on a literal or an imaginative level, might be effectively invited to participate in the characters' emotions.

5. The sentimental plot emphasizes close calls and last-minute reversals, either for better or for worse. Charlotte drops dead only moments after her father finally finds her; young George Shelby arrives just too late to save Tom; Eliza escapes over the frozen Ohio River with no time to spare. The reader who allows him or herself to

become absorbed in the fictional world can experience emotional jolts when expectations are dashed or suddenly fulfilled, and those jolts can bring on tears.

6. Contrary to popular belief, sentimental texts do not present two-dimensionally "stereotyped" characters, but rather rely on characterization that mixes traits to work against the types that have been established in the text (or in the culture). This strategy works together with the unexpected reversals of the plots to prod a reader's emotions. An example of such an exploded character type would be the respectable, emotionally repressed, publicly powerful middle-class patriarch—like Senator Bird in *Uncle Tom* or Mr. Laurence in *Little Women*—who turns out, against his type, to be capable of forming an emotional bond with a slave or a child.

Consider, for example, the scene in which Beth astonishes her family by overcoming her shyness (up to this point, her dominant trait) and visits Mr. Laurence to thank him for the piano he has sent her:

> If you will believe me, she went and knocked at the study door before she gave herself time to think, and when a gruff voice called out, "Come in!" she did go in right up to Mr. Laurence, who looked quite taken aback, and held out her hand, saying, with only a small quaver in her voice, "I came to thank you, sir, for—" But she didn't finish, for he looked so friendly that she forgot her speech and, only remembering that he had lost the little girl he loved, she put both arms round his neck and kissed him.
>
> If the roof of the house had suddenly flown off, the old gentlemen wouldn't have been more astonished; but he liked it—oh, dear, yes, he liked it amazingly!—and was so touched and pleased by that confiding little kiss that all his crustiness vanished; and he just set her on his knee, and laid his wrinkled cheek against her rosy one, feeling as if he had got his own little granddaughter back again. (60)

Here, both Beth and her benefactor behave against type, and the little narrative ejaculation ("oh, dear, yes!") adds the stylistic prompt for the reader's emotional response.

7. The plots of sentimental novels rely upon repeated moments, like this scene between Beth and Mr. Laurence, which serve to "mythologize" experience in that they draw upon the culture's store of cherished beliefs and—contrary to the perhaps painful evi-

dence of readers' own lived experience—make those beliefs seem to come true. In cases where the subject matter is as politically charged as it is in *Uncle Tom*, this can mean reducing situations of extreme moral and logistical complexity, if only momentarily, to epiphanies of pure simplicity. This last, and most powerful of the techniques of sentimental narrative discourse can account—as Fisher's model cannot—for the "tears of triumph" that dominate *Uncle Tom* and *Little Women*, as well as the "tears of defeat" that appear to have prevailed among the readers of *Charlotte Temple*.[11] When—in *Little Women*—Mr. Laurence's dead granddaughter seems to have returned to him in the form of Beth, when Meg and Jo are reconciled after a long period of bickering, when Laurie and Amy become engaged; or—in *Uncle Tom's Cabin*—when Eliza and George are reunited at the Quaker settlement on their journey to Canada, when the bigoted Vermonter Ophelia finally comes to love the slave Topsy, when Eva or Uncle Tom dies with the most confident expectation of going to heaven, the reader who weeps is probably experiencing something other than helpless frustration. This reader is having a "good cry," an affirmation of the mythology being represented in the text: family affection *does* transcend death; sisters *are* friends forever; true love *will* prevail; courage *will* be rewarded; Christian forebearance *will* put an end to slavery—oh, it *is* a wonderful life!

I can't do it without sarcasm; the legacy of modernist irony makes that impossible. The association of the feelings with bourgeois mythologies makes them suspect, false, "sentimental." But from the nineteenth-century perspective, the weeping reader, whose body is experiencing the signs of an emotion, is experiencing the genuine feeling itself, no matter how "mechanically" it may have been produced. The experience is not a cathartic one: the weeping does not purge an emotion, but rather produces and reproduces it in the responsive reader, and thus very effectively constructs the novel's audience. I think our academically sanctioned annoyance and embarrassment at being "made to cry" is a specifically modern reaction (and therefore anachronistic in a postmodern age), analogous to our assumption that an actor must experience a "real feeling" first, in order to express that feeling through physical gesture. When Stowe and Alcott were writing, the assumption was rather that since the emotion originated in the gesture, the reader who could be made to cry could literally *realize* the message of Christian domesticity, the mythology that informs and motivates the novels' narrative strategies. For such novelists, to make a reader cry would be to make something real happen.

And that something continues to happen to susceptible readers, whether they are encountering *Little Women* or this century's versions of those domestic mythologies, most often taking the form of melodramatic films or television programs. In the twentieth century, crying—under any but the most extreme circumstances of personal triumph or defeat—is coded as feminine, childish, something for gushing girls to do. I think it cannot be a coincidence that this association has developed at the same time that those texts which induce crying have been systematically devalued by critics in the academy and in the popular press. If we could put aside the prejudice against "having a good cry" over a book or a film, we would be in a position to envision a truly inclusive cultural poetics, where the evaluative standards of past eras might no longer have the power to exclude difference and diversity in descriptions of texts, or of the audiences who receive them. To stop thinking of catharsis as an essential or natural explanation for the relation between the body and the text would be a positive step in that direction.

And what about the sentimental audiences of the twentieth century, those persons (male or female) who, perhaps partly unwillingly, can "choke and cry" the feminine tears evoked by these machineries? I have sketched the beginnings of a list of mechanisms American texts use to make audiences cry, but I do not offer this list to those audiences as a weapon against the sentimental effect. Like therapy, it does not work that way—even if you understand how the process is supposed to operate, that does not guarantee you will be impervious to its effects. Creatures of culture that we are, we may not be in the position to change the emotional reactions that have been programmed into us. But what we do with the reaction, and what sort of attitude we take toward it, *can* change. We don't have to be embarrassed when our tears remind us that we are living in bodies, and that we are repeating the experience of generations of gushing girls before us. If the reservoir/catharsis model has indeed led to a view of crying audiences as alienated, isolated, and disempowered, I suggest we take a cue from the Delsartean model, which saw emotion as publicly, even communally produced. We have long been accustomed to thinking of audiences who laugh at comic productions as constituting a kind of community, finding a form of power in their common experience. Why should laughter be regarded as so essentially different from crying, just another bodily function associated with an emotion? I propose we envision a community of readers empowered by their relationship to sentimental texts, fully conscious of how strategies "get to them," and free to enjoy having a good cry. Once that community's sense of embarrassment and isolation has been dispelled, its potential for participating in the transformation of culture and society will be considerable.[12]

NOTES

1. Jack Kroll.

2. Positive attention to nineteenth-century sentimental novels has been gradually building upon the groundbreaking work Helen Papashvily did in the 1950s. Ann Douglas subjected sentimentalism to a serious political critique in 1977, but more recently such critics as Nina Baym, Mary Kelley, Jane Tompkins, and Philip Fisher have been demonstrating a willingness to take the genre on its own terms.

3. Marshall Fine for Gannett News Service.

4. Judith Newton locates the emerging "suppression of feeling" among men in the 1860s and 1870s, "in the shape of masculine ideals which emphasize separation from women rather than appropriation of their virtues"; the ideals "devalue feeling and domesticity and valorize 'manliness' defined as 'anti-effeminacy, stiff-upper-lippery, and physical hardiness'" (134–135). Like Newton, I do not mean to claim crying as in any sense an essentially "female" activity; when I say it is "coded feminine," I mean that femininity is the mark of culture upon the gesture.

5. My collaborator, Jane Shattuc, and I are at present designing a study to redress this lack.

6. Tompkins's battle against modernist bias is especially evident in "An Introduction to Reader-Response Criticism." As her anthology shows, Tompkins is not alone in avoiding discussion of readers' physical experience. For all reader-response theorists to date, the "reader" is a construct, primarily useful in projects whose ends are interpretive. As Steven Mailloux, for instance, explains, "Functionally within my discourse, 'the reader' is an interpretive device for literary theory, practical criticism, textual scholarship, and literary history" (192); in such studies, the reader's body is beside the point. Indeed, the only work of criticism I have seen that directs its attention to the interactions of a text with a reader's body is D. A. Miller's recent virtuoso piece about sensation fiction, *"Cage aux folles."*

7. Lewes's periodical articles on the theater are collected in *On Actors and the Art of Acting.*

8. Banta focuses her discussion of Delsarte's techniques on the mean-

ings implied by the poses, rather than the emotional states supposed to be brought on by them.

9. Considering its source, Davidson's treatment of the reading process is peculiarly ahistorical. According to Davidson, "Within fictional space-time, the ahistorical out-of-time dimension of the reading process, there occurs a transubstantiation wherein the word becomes flesh, the text becomes the reader, the reader becomes the hero." Davidson attributes *Charlotte Temple*'s popularity to this "intimate, transformative process of reading." Having taught the novel to undergraduates, however, I question its ability to transcend historical differences between young readers in the 1790s and the 1990s. Davidson, famous for historicizing texts, demonstrates here the propensity of many critics for not historicizing the reading process.

10. See the introduction to *Gendered Interventions*.

11. Davidson's account of Rowson's readership suggests that those who cried saw Charlotte as a helpless victim, her fate "not as justice but as a tragic metaphor for human pain—for the reader's pain—in all its variety" (170). I borrow the distinction between tears of "triumph" and of "defeat" from James L. Smith, who anatomizes nineteenth-century melodramas in both modes.

12. I want to thank my own community of susceptible readers for their helpful comments on this essay, especially Diane Price Herndl, Beth Kowaleski-Wallace, Roxanne Lin, and Helena Michie.

WORKS CITED

Alcott, Louisa May. *Little Women.* 1868–1869. Ed. Nina Auerbach. New York: Bantam, 1983.

Aristotle. "The Poetics." ca. 335–322 B.C. *Criticism: The Major Statements.* Second edition. Ed. Charles Kaplan. New York: St. Martin's, 1986. 21–53.

Banta, Martha. *Imaging American Women: Idea and Ideals in Cultural History.* New York: Columbia University Press, 1987.

Baym, Nina. *Novels, Readers, and Reviewers: Responses to Fiction in Antebellum America.* Ithaca: Cornell University Press, 1984.

———. *Woman's Fiction: A Guide to Novels by and about Women in America, 1820–1870.* Ithaca: Cornell University Press, 1978.

Breuer, Josef and Sigmund Freud. *Studies on Hysteria (1893–1895).* Trans. James Strachey. New York: Basic, 1957.

Davidson, Cathy N. "The Life and Times of *Charlotte Temple:* the Biography of a Book." *Reading in America: Literature and Social History.* Baltimore: Johns Hopkins University Press, 1989. 157–179.

Douglas, Ann. *The Feminization of American Culture.* New York: Knopf, 1977.

Fine, Marshall. "'Stella' Really Lathers up the Soap." *Burlington Free Press.* (3 February 1990): 5A.

Fisher, Philip. *Hard Facts: Setting and Form in the American Novel.* New York: Oxford University Press, 1987.

Hamilton, Alice. *Exploring the Dangerous Trades.* Boston: Little, Brown, 1943.

Hinds, Julie. "Get Out Your Handkerchiefs." *Burlington Free Press* (30 November 1989): 13D.

Kelley, Mary. *Private Woman, Public Stage.* New York: Oxford University Press, 1984.

Kroll, Jack. "A 'Winter's Tale' as You Like It: Shakespeare's Alive and Well and in Bohemia." *Newsweek* (3 April 1989): 70.

Lewes, George Henry. *On Actors and the Art of Acting.* New York: Holt, 1892.

Mailloux, Steven. *Interpretive Conventions: The Reader in the Study of American Fiction.* Ithaca: Cornell University Press, 1982.

Miller, D. A. "*Cage aux folles:* Sensation and Gender in Wilkie Collins's *The Woman in White.*" *Speaking of Gender.* Ed. Elaine Showalter. New York: Routledge, 1989.

Newton, Judith. "Making—and Remaking—History: Another Look at 'Patriarchy'." *Feminist Issues in Literary Scholarship.* Ed. Shari Benstock. Bloomington: Indiana University Press, 1987. 124–140.

Papashvily, Helen. *All the Happy Endings: A Study of the Domestic Novel in America: the Women who Wrote it, the Women who Read it, in the 19th Century.* New York: Harper, 1956.

Roach, Joseph. *The Player's Passion: Studies in the Science of Acting.* Newark: University of Delaware Press, 1985.

Rowson, Susanna. *Charlotte Temple.* 1794. Ed. Cathy N. Davidson. New York: Oxford University Press, 1986.

Sicherman, Barbara. "Sense and Sensibility: A Case Study of Women's Reading in Late-Victorian America." *Reading in America: Literature and*

Social History. Ed. Cathy N. Davidson. Baltimore: Johns Hopkins University Press, 1989. 201–225.

Smith, Herbert F. *The Popular American Novel: 1865–1920.* Boston: Twayne, 1980.

Smith, James L. *Melodrama. The Critical Idiom 28.* General editor John D. Jump. London: Methuen, 1973.

Stebbins, Genevieve. *Delsarte System of Expression.* Sixth ed. (1902). New York: Dance Horizons, 1977.

Stowe, Harriet Beecher. *Uncle Tom's Cabin.* 1852. Ed. Ann Douglas. Harmondsworth: Penguin, 1981.

Tavris, Carol. *Anger: The Misunderstood Emotion.* New York: Simon and Schuster, 1982.

Taylor, Helen. *Scarlett's Women: Gone with the Wind and its Female Fans.* New Brunswick: Rutgers University Press, 1989.

Todd, Janet. *Sensibility: An Introduction.* London: Methuen, 1986.

Tompkins, Jane. "Introduction to Reader-Response Criticism." *Reader-Response Criticism: From Formalism to Post-Structuralism.* Baltimore: Johns Hopkins University Press, 1980. ix–xxvi.

———. "Me and My Shadow." *Gender & Theory: Dialogues on Feminist Criticism.* Ed. Linda Kauffman. Oxford: Basil Blackwell, 1989.

———. *Sensational Designs: the Cultural Work of American Fiction: 1790–1860.* New York: Oxford University Press, 1985.

Warhol, Robyn R. *Gendered Interventions: Narrative Discourse in the Victorian Novel.* New Brunswick: Rutgers University Press, 1989.

7

Torture: A Discourse
on Practice

In September 1984, the Comisión Naciónal sobre la Desapari-
cion de las Personas (CONADEP), the Argentine National Committee on the Disappeared, submitted to President Raúl Alfonsín a final
report on the thousands of cases of torture that had taken place in
the jails and concentration camps of Argentina between 1976 and
1979. The report was soon published under the title *Nunca Más*
(Never Again), and became an international best seller. In giving to
the press the results of their painstaking research, the members of
the Committee confess not only to their sadness and pain before the
"dreary puzzle" of the "greatest and most savage tragedy in Argentine history," but also to their patriotic embarrassment in verifying
that Argentina had become an example of barbarism and political
backwardness in the eyes of the "civilized nations." It is, they say, "a
sad Argentine privilege," that today the international press writes
the word "desaparecido" in Spanish. The preface to the Report concludes with an exhortation to all Argentineans to fight for the con-

126

solidation of democracy, the only political strategy that, according to them, will ensure that "never again the facts that have made our country tragically famous in the civilized world will be repeated."

Recurrent images borrowed from the language of theater and drama heighten the emotional tone of the report, which describes the events in Argentina as a barbarous "tragedy" unfolding before "the eyes of the civilized world." Obviously, the members of the Committee see Argentina as a mostly educated, essentially democratic, ultimately "civilized" nation. For them, indeed, the "dreary puzzle of the Argentine nationality," the truly tragic aspect of the problem, is that, during the years of military rule, the country gave an image of itself which did not correspond to the liberal traditions, the nonviolent temperament, and the cosmopolitan makeup of its population. How it is possible then, they seem to ask themselves, that Argentina has come to symbolize what it is not?

Author and politician Domingo Faustino Sarmiento had asked the same question over a hundred years earlier in his *Facundo: Civilización y barbarie*, one of the classics of Argentine literature. The violent instability, the bloodthirsty arbitrariness of power, supposedly typical of Latin American politics, was also a source of shame for the enlightened, European-minded author of *Facundo*. Argentina is a rich, unpopulated country, he thought, without any serious historical traumas or any unsolvable social conflicts. Why is it then that it always seems to be falling short of its own bright promises? The authors of *Nunca Más* express a similar sense of puzzlement in concluding that there is no convincing explanation for the unchecked reign of brute force in a society that they perceive as part of the modern, educated Western world. For them, torture is incomprehensible, unjustified, alien, like the obstinate trace of an abjection that the civilized eye can only apprehend from the outside and for the outside in the civilized body.

This "dramatic" perspective, common among Argentine thinkers, who see themselves as powerless spectators in an exemplary "theater of cruelty," is also prevalent in European and North American political circles. In the United States, even among liberal observers, compassion for the victims of torture is almost invariably accompanied by an ethnocentric contempt for the societies where torture is practiced, which are thought to be imperfect, immature political systems, prone to fluctuate between anarchy and authoritarianism. Only Anglo-Saxon style democracy, the argument goes, can protect the Latin people from their own innate unruliness.

Of course, to the left of the political spectrum this patronizing view is mitigated by a more responsible, often even self-incriminating conviction that the ruthless dynamics of advanced capitalism in the so-called First World makes economic and political injustice unavoidable in the so-called Third World. Freedom and affluence in the North, it is thought, are contingent upon oppression and poverty in the South. This perspective is generally shared by the North American liberals, the European left and the Latin American *intelligentsia*. For the sake of explanation, I will refer to it as the "global perspective."

According to the global perspective, totalitarianism in the Third World is not a peripheral, outdated political phenomenon, but rather an integral part of a global system that concentrates power and resources in some countries by depriving other countries of their own wealth and autonomy. It is undeniable that this contention is backed up by overwhelming evidence of the support that the governments of the "developed" countries have lent and are still lending to totalitarian regimes in the "Third World." It is also undeniable that the "global attitude" is inspired by a sincere solidarity with the cause of the oppressed, and that it often leads to concrete political activism against totalitarianism. I still think that its solidarity is often colored by a vague paternalism, and that so far the results of its activism have been episodic and only palliative.

The purpose of this paper is to confute the premises of the global attitude, to question its theory of power, and to controvert the understanding of the human body that it implies. It is my conviction that as soon as we start studying torture as the reflection of distant, centralized power, we are immediately forced to resort to a series of logical polarities, such as "center" versus "periphery" and "strong" versus "weak," which are misguiding and detrimental to a truly sympathetic view of the Latin American people. The fortitude with which the Argentine people resisted the unrelenting threat of torture indicates that a characterization of them as weak is inadequate. To define those societies as culturally peripheral is equally misleading. A European oriented education has instilled in most Argentineans an alienating displacement of perspective, a sensation that "the center of culture," that "what really counts," that "reality," in sum, always takes place elsewhere. But how can we accept our own marginality when torture itself reminds us that our center, our reality takes place inside our society and needs to be confronted there?

In a long footnote to his *Speed and Politics*, Paul Virilio has explained the transition from institutionalized torture to the modern, supposedly "benevolent," penal system, which took place at the end of the eighteenth century, as the result of "a devaluation of confession." In the Middle Ages, Virilio says, the question was imposed under torture onto a body which was assumed to know the truth. According to him, it was not out of humane concerns that torture was abolished in the nineteenth century, but rather because it became evident that the body, in spite of itself, leaves behind traces, involuntary evidence, proofs (40). Judicial inquiry became then an obsessive quest for signs of the inattention, the negligence, the oversights of a body that had not been totally subject to grooming, to drilling, to discipline.

Virilio borrows from Michel Foucault an already classical distinction between "monarchical power" and "disciplinary power." According to Foucault, premodern monarchy derived its power from its own visibility, dramatically unfolded in scenes of public punishment. The power of the disciplinary state, by contrast, is hidden from its subjects. A constant but imperceptible watch over the citizens normalizes public behavior even before deviance arises and turns punishment into an only auxiliary tool of power. In the modern hospital, in the modern school, as in the modern prison, each person is a compliant actor performing in front of an invisible audience.[1] Monarchical power manifests itself on the body of the criminal, and its exercise resides in punishment. The point of application of disciplinary power, on the other hand, is not exactly the body, but rather the image, the projection of the body of the "citizen," (no longer the criminal), and its exercise consists in the judicial investigation. The agent of justice is no longer the executioner, but the detective. The magnifying glass has taken the place of the stake.

But what about contemporary torture? Virilio's distinction seems to ignore that in the twentieth century the majority of the world population still lives under regimes that practice it. Is he blinded by the Eurocentric scope of Foucault's distinction, which is probably appropriate to describe France and other European societies, but fails when matched with information coming from other, dictatorial, mostly non-European societies? It is ironic that, at a time when torture has disappeared from the developed societies, its persistence in other societies has become a criterion for classifying them as "underdeveloped" or "premodern." In Europe and the United States, torture, rejected as a political practice, is still emphasized as an instrument of political knowledge. Dislodged as a judicial entity, it has survived as an epistemological construct. There, pain is no longer an element of political reality; instead, it has become a standard for classifying nations.

Paradoxically, societies that no longer practice torture acknowledge it as part of their own political representation. In the Argentine society, by contrast, where torture is still a concrete menace, the educated classes insist on denying its function in the general structure of power. The final report of the CONADEP, for example, suggests that democratic fair play will soon make clear to all Argentineans that political violence is alien to the truly modern, democratic, Western spirit of their nation. Faithful to Sarmiento's humanist tradition, the compilers cling to an image of their country as a "European" society. Since the time of Independence, in 1810, most intellectuals have accepted such an image, propped by the cosmopolitan, mostly urban makeup of the Argentine population. A few, however, have challenged it. Juan Bautista Alberdi, for example, insisted that a specifically American, primarily violent streak was at work in the Argentine character, and accused his contemporary Sarmiento of intellectual snobbery for refusing to perceive it. Nevertheless, it would be unfair of us today to extend that accusation to the compilers of *Nunca Más*. Their reluctance to accept violence as an unavoidable component of Argentine politics is certainly not based on alienated snobbery. It is, above all, the result of a mediated decision before an exacting ethical dilemma: How to explain torture without justifying it, without rationalizing it, without excusing it?

In a sequence of the film *The Battle of Algiers*, by Gillo Pontecorvo, the general in charge of the French occupation forces explains to the foreign press that "the specific conditions of the Algerian situation" have forced him to take "special measures." The Argentine generals seem to have learned this lesson in foreign cynicism. Although they never admitted that there were any concentration camps or torture chambers in Argentina, they always complained about the lack of understanding, the blindness, the hostility of the foreign press. They insisted that foreign observers failed to identify the special conditions that made strong measures necessary in Argentina, and tried to press themselves as the tragically misunderstood defenders of democracy against communism.

For the members of the CONADEP, the suppression of civil liberties was a "remnant of the Dark Ages." For the military, on the contrary, it was the painful but necessary price that Argentina had to pay in the defense of democracy not only inside its borders, but also in the world at large. Their convoluted reasoning, which obviously departed from the language of "weakness" and periphery implicit in the "global" hypothesis, pretended that their reaction against the communist threat was at once specific, since it was adapted to the native conditions, and global, since it was devised as a defense of transnational values. "Strong measures" were the sacrifice that modern democracy demanded from its most vigilant defenders.

The exculpation was essentially lucid, insofar as it acknowledged the need for understanding torture from the inside, from the conditions that make it possible. But it was also cynical, because, while defining the appropriate perspective, it insisted on denying the existence of the object. "We could explain torture . . . , the Generals seemed to promise, but we don't have to, since there is no torture in Argentina." They denied and justified torture at the same time. Or, more exactly, they justified it by denying its existence.

We can read Horacio Mino Retamozo's story among the innumerable gruesome accounts of *Nunca Más* (29–33). He was kidnapped on August 23, 1976, in Buenos Aires. They blindfolded him and "caressed" (that is his own word) his naked body with electric shocks. A medical doctor monitored the operation, lest Retamozo faint or die before speaking. They asked him about *"el cope del rim"*; he protested he did not even know those words. Exasperated, they tied him to the "grill" (as they had named a metal bed frame) and tortured him even more. Then, they changed the meaning and the formulation of the questions, but the essence of the charges persisted. He did not react, since he was unable to understand the words in which the accusation was formulated. His very innocence seemed to madden them even more. Now, they introduced an electric device through his throat and gave him an internal shock. He remembers now that the inability to scream made his pain indescribably intense. Their violence became more frenzied: the famous "feminine voice," a particularly vicious officer who imitated a woman, caressed his testicles "anticipating the enjoyment of his work." They put an electric helmet on his head; when he got the shocks, he could not even say "no."

After ten months of captivity, another prisoner confided to Retamozo that the expression *"el cope del rim"* referred to *"el copamiento del Regimiento de Infantería de Monte número 29,"* a terrorist attack on a military base in Formosa, in Northeastern Argentina. Shortly afterwards, another prisoner, who was in a cell across the corridor from his, managed to tell him that a certain Mirta Infrán, a classmate with whom Retamozo was linked by a "tangential friendship," had incriminated him. Her husband had been tortured to death; when her turn came, "she went astray, she either tried to save herself or stumbled over the threshold of madness," as the text puts it, and started to "sing . . . unbelievable things." Retamozo, in the erratic logic of her delirium, had become one of the leaders of the attack.

Retamozo was finally released on June 6, 1977. Upon leaving the prison camp, he got from his captors the half-apologetic, half-ironic admission that somebody had made "a mistake." "In the middle of this tragedy, absurdity," comments the CONADEP. A man "who did not even understand what he was being asked" had been mistreated to the limits of physical and psychological tolerance. Indeed, the poignancy of Retamozo's account comes to a great extent from the sensation of painful absurdity that it suggests to the reader. But, where does this sensation itself come from? What exactly is it that is absurd, baffling, irrational about this account? Certainly not the innocence of the accused. As a matter of fact, the very notion of "innocence" is irrelevant from my point of view, in part because I do not believe that resisting a totalitarian regime is a crime, but primarily because I think torture is immoral even when the accused is indisputably guilty.

The testimony tells us that Mirta Infrán's "delirium" had sent fifty innocent people to the torture chambers. She made several mistakes in giving the names of innocent people; the torturers then made several mistakes in persecuting those innocent people. . . . There can even be a certain logic to the nonsense of torture. It is repeated, consistent, almost predictable, never exactly "absurd." Indeed, it would have been outrageous if it had never occurred to any of Mirta Infrán's torturers that she was not telling the truth. The sensation of absurdity, in Retamozo's account as in many other accounts, does not arise from the inability of torture to produce truth, but rather from its consistent disregard for it. A reader with even a superficial knowledge of the Latin American political reality is never surprised by the dreadful effects of torture. What puzzles, what shocks, is instead the uncanny consistency of those effects with a purpose so different from the one torture claims to have. What one finds absurd is the cynicism of its rationality.

In a story by Elvira Orphée the fictional narrator, an Argentine policeman, admits that he is more suspicious of his victims when they talk under torture than when they keep silent. Excruciating pain, he says, may drive somebody to accept that he is guilty and even to incriminate other innocent people (48). Many direct, nonfictional accounts confirm that the executioners think that *ideally* torture is a way to find out the truth, but that they also admit that the *practice* of torture gives rise to many distortions. A torturer is aware of the dangerous side effects of his work, but he still believes that they could be avoided. His dream is a kind a surgical torment, localized, without any side effects, whose properties could be separated from its essence, like one of those foul smelling substances that the makers of perfume use to give persistence to their fragrances. Pain would then become its own double, forever threatening the disciplined body of the modern person, a ghost that chases an appearance. Such is the Fascist utopia: an immaterial, truly incorporeal encounter between the idea of torture and the projection of the body, a society without a single individual, but with only one truth. Horrors without errors.

Of course, some torture for pleasure. They are probably cruel but shy, and they become policemen the way some pyromaniacs become firemen. They are not many, though. Fascism could hardly prosper on a principle as potentially subversive as pleasure, even if it were the most abject of pleasures. Instead, everything seems to indicate that the typical torturer sincerely believes in his job as a means to investigate the truth. At the same time, however, he knows that the victim can be innocent and that the information that he gives is not always accurate. And yet, the contradiction seems not to affect the torturer. His preoccupation is different. Actually, it does not matter to him what the victim says; the only thing that matters is that he talk. *El cope del rim?* Yes, I am guilty. Anything. I am guilty of anything, everybody is guilty of everything. Torture is the realm of the pure signifier. Or even better: torture is the *democracy* of the pure signifier. Once the signified is withdrawn, anything is equal to anything. As there is no truth, there are no mistakes; there are merely errors. And torture is the production of the signified as illusion, as absence, as *errancy.*

Language, the body, truth: the formula seems irresistible to the followers of what in the United States is loosely called "postmodernism." Over the last decade, bookstore shelves have been inundated by anthologies, collections of articles, and exegetic commentaries that promise to elucidate for the lay person this new trend that, we are told, has revolutionized academic "discourse." The apostles of the new cult are almost invariably French: Foucault, Derrida, Lacan, Baudrillard. The commentators are instead almost always North Americans. With the exception of Baudrillard, none of these French authors would recognize himself behind the label, and very few, again with the exception of Baudrillard, has written extensively on the subject. My paper, therefore, does not exactly allude to these writers that only simplistically could be termed "postmodern." It refers, instead, to the version of their work that we receive from the North American publishing industry and the academic rhetoric. Postmodernism, as I treat it in this paper, is a "vogue," a "trend," in the sense of something consumed by people who are "trendy."[2] As a trend, postmodernism is exempt from all expository consistency and all organizational support. It is an aggregate of bibliographical references, linguistic turns, ideological poses, artistic tastes.

A trend is not exactly a discourse, or an institution; it is mere "effect." Before its elusiveness, criticism hesitates between a enumerative, statistical approach (which sacrifices the impression, the very effect that defines the object), and a phenomenological interpretation (which can potentially remain on the psychological surface of the effect, leaving its political dimensions unexplored). The solution, whose definitive form I will have to postpone for a future paper, seems to lie in an approach at the same time *rhetorical*, that would describe the figures of postmodernism ("language," "the body," "truth," for example), *genealogical*, which would explain how those figures derive their legitimacy and prestige from certain cultural theories, and, finally, *political*, which would recount the different dispensations of power made possible by those figures.

The distinction that I have drawn here between the "theory" of some contemporary French thinkers and the trend of some mostly North American circles should not be interpreted as yet another attempt to protect the presumed sophistication of Continental thought from the supposedly naive simplifications of New World pragmatism. On the contrary, I have concentrated on the trend precisely because I think it is a field in which theory, free from the constraints of logical consistency and the mystique of academic seclusion, opens itself to innovation and dissemination. Half way between scholarly research and mass culture, the trend, as a typically "middle-brow" phenomenon, is the field in which theory is simplified in order to be redirected, circulated, utilized.

In its political utilization, postmodernism is independent from any philosophical system. Strictly speaking, the philosophy to which it alludes constantly is not a system, insofar as its propounders represent different, often opposed views. If anything, it is postmodernism itself that assorts them together into a fairly consistent group, from which the formulations of the trend derive their language, their style, and their legitimacy. In this other sense, which I have called genealogical, postmodernism is deeply indebted to European academic thought, and it could not be understood fully without following the main developments in the disciplines that concerned themselves with language, the body, and truth.

At the beginning of the twentieth century, Ferdinand de Saussure revolutionized linguistics, and eventually all the other social sciences, by advocating that language should be studied not in reference to *what* it signifies (reality, truth, the world) but rather in reference to *how* it signifies. Saussure posited that linguistic laws are independent from the degree of accuracy of the discourse. As Plato had suggested in his "Cratylos" (385), people use the same language to lie and to tell the truth. Those laws, continued Saussure, should be studied as a self sufficient, self-contained order, as a structure (a system, that is, equally independent from nature, in which they were nonetheless rooted, and from reason, to which, Saussure acknowledged, they were inextricable connected).

The concept of "structure," which evoked notions of quantitative knowledge and mechanical efficiency, was particularly fitting to describe the phonological method, whose object could be measured with relative ease. Nevertheless, as the Saussurean revolution reached other fields where questions of precision and efficacy were more conspicuously moot, the term structure started to be perceived as somewhat inappropriate. The human sciences, after a short-lived fascination with the aura of hard exactness and rigor that surrounded the notion of structure, found in their own conception of the human body a more adequate metaphor to explain their endeavors. For example, post-structural anthropological theory, which claims to have ridded itself of any illusions of mathematical rigor, resorts now to the image of the body with suspicious insistence.

The body, that half-poetic, half-obscene, always elusive concept, has become almost *de rigueur* in the title of any anthropological work. Pre-structural anthropology was based on an organicistic idea of culture: in the specific, extreme example of Spencer, societies were born, developed and died, and went through something like a life cycle thanks to certain institutions which, like organs, performed certain functions so that the embryonic design of history and the natural demands of the environment were satisfied. Post-Saussurean anthropology has dispensed with the notion of an embryonic design, and finds no need to refer the development of culture to any natural, that is to say extra-cultural, demands. In other words, it has substituted the body for the "organism."

It should be clear that this distinction between organism and body, which contemporary theory uses as a metaphor to define both its method and its object, does not come from biology, but rather from anthropology itself. While medicine and biology concerned themselves with the organism, anthropology became interested in the body, a hybrid entity that supposedly reflected the ambivalence of human experience, born of nature and shaped by culture. A method, an object, a metaphor for the knowledge of man, the body, as understood by contemporary theory, represents a self-sufficient, self-regulated structure, a "plan of organization" that is at once autonomous from nature and truth, only subject to its own logic (Deleuze and Parnet 110).

The object of social inquiry is never independent from the theoretical model that defines its limits, enunciates its laws, and in a general sense, governs its structural consistency. Conversely, knowledge is unavoidably immersed in the logic of its object, and must therefore start by an appraisal of the place of the observer in relationship to that object. That "place" is at once the regulating principle of the structure, since it defines it, and the point that most intensely resists the advances of the inquiry, since it determines, and therefore precedes, the interest and the method of the observer. That empty place, that blind spot, that moot point where the mind constructs the world and the world dissolves into the mind is the place of the anthropological body.

From a certain perspective, the anthropological notion of the body seems to transcend the dichotomy between consciousness and things, between representation and its *representatum*, upon which Western logic was based since the Greeks. From another, more attentive perspective, however, it makes patent the difficulties of transcending that dichotomy without falling into a solipsistic trap. In the past, in a more self-confident mood, anthropology viewed the body as the "natural" component of man, as that which in man was not anthropological. But now, in a moment of crisis which imposes a certain introspective mood in its pursuits, the discipline insists on defining its object and its method as a body, thus resorting to a word it had originally excluded from its vocabulary. Nevertheless, the resurgence of the term does not totally efface the traditional line between nature and culture. In a sense, it draws it even more sharply. Ironically, almost perversely, anthropology represents itself as the body, not as the anatomical body, the organism, but as the body as defined by anthropology. Circularly, in nearly autistic fashion, the knowledge of man becomes a metaphor for itself, an empty mirror, an echo. The body, which had promised to mend a cognitive fissure, has only succeeded in hiding a narcissistic dead-end.

In the realm of political theory, by contrast, we saw that the study of torture produced a different epistemology, in which the distinction between object and representation survived in spite of the assault (or under the shelter?) of postmodern jargon. Torture has become the spoils in this tug-of-war between a discipline that seeks in vain to heal the wound that lies between humans and things and another discipline that obstinately insists in laying it bare. Anthropologically, in the limits of nature, the body is the result of a harmony with design; politically, in the center of culture, the same body appears as an intrusion into the order of the world.

"To speak is above all to possess the power to speak. Or again, the exercise of power ensures the domination of speech. Only the masters can speak," writes Pierre Clastres in his *Society Against the State* (133). Félix Guattari and Gilles Deleuze, discussing Clastres' findings about torture in primitive societies, conclude that its purpose is to "mark" people, to give them a place in the social structure, to ensure that everyone is equal to everyone in relationship with the presence of power (163–170). Clastres defines power as an *active* force operating among the people, as something *present* among them through the *visible* mark it has produced on their bodies. The practice of torture in the Westernized societies of Latin America indicates a connection between language and power which is exactly opposite to the one described by Clastres. In Argentina, in the 1970s, power was the power not to speak. The purpose of contemporary torture is not to assign a place to each human being, but rather to break him or her down, to make the "subject" *cease to be*, as one of the informants to the CONADEP puts it.

Elaine Scarry, in a very influential book about the political utilization of physical pain, has suggested that the effect, if not the purpose, of torture is to move the victim from the realm of interpersonal communication back to a pre-linguistic stage in which defenseless crying, inarticulate screaming, primordial shouting, signal the destruction of language and everything that, according to her, language stands for: meaning, understanding, civilization:

> To witness the moment when pain causes a reversion to the pre-language of cries and groans is to witness the destruction of language; but conversely, to be present when a person moves up out of that language and projects the facts of sentience into speech is almost to have been present at the birth of language itself. (6)

Scarry, who in the end conforms to an organicistic, almost embryonic model of culture, contends that pain drives the individual back to a stage of psychological development in which language is not yet articulated. She implies that the ultimate outcome of torture is to separate the individual from the ability to produce meaning. The body would then revert to an emotional experience similar to that of children and to the mentality of archaic societies. The destruction of civilization would thus uncover what supposedly was there before civilization, its fount and its substratum: primitivism, childhood, the primal scream, the lack of articulation, the absence of meaning.

And yet, what makes Mirta Infrán's confession valuable for her torturers is that, *although it is not accurate,* it does make sense, it does convey a message, it does after all have a political effect. Pain has forced her to give concrete (but false) names, exact (but random) addresses, precise (but distorted) information. Her language has not regressed to a pre-coded, inarticulate, wild scream. On the contrary: it has proceeded to a point where its subjection to the code, its conformity with syntax, its artificiality, has allowed it to perpetuate itself, regardless of its accuracy. Mirta Infrán's confession is perfectly articulated, but also perfectly arbitrary. Contrary to what Scarry would have argued, it comes from a body which is still (or already) controlled by the laws of language. Pain, the pain inflicted by contemporary torture, does not break down a pre-existing subject. It does something more and something less than that: paradoxically, it produces the subject as already (or still) absent.

As Elaine Scarry has pointed out, people who speak under torture are often ostracized by political activists and generally slighted by the literature about torture. Those who have never suffered intense physical pain seem to believe that it is always possible to resist it. In their mind, succumbing to torture is always the result of a cowardly predisposition or of a latent propensity to collaborate with the tormenters. The members of the CONADEP avoid taking this sanctimoniously comfortable position in discussing Mirta Infrán's case. Driven to the extremes of pain and humiliation, they write, she has "stumbled over the threshold of madness," and suggest that she cannot be held responsible for her mistakes because she is no longer herself.

In few words: Mirta Infrán does not tell the truth because she is no longer herself. The premises of the exculpation are obviously grounded on a metaphysical understanding of language and consciousness, of which contemporary thought, and postmodernism in particular, has taught us to be suspicious. I still think that they reflect a very lucid perception of the effects of torture, and will not directly challenge them here. The threat of pain does link in the mind of the victim the experience of the self and the recognition of truth, but only through their negation. Pain does not resuscitate the Cartesian plenitude of the subject apprehending reality. It merely homologates the absence of truth with the absence of the self. From this perspective, we can now define the "work" of torture as *counterproduction*, in the sense that it is *a kind of work whose product is preceded by a negative sign*.

Scarry has argued that torture destroys language and the self; I am contending, instead, that it produces them as negative, as absent, as disembodied images. I do not think that pain substracts the victim from the realm of articulated language; I think, instead, that torture forces pain to become something like a "transcendental signifier," the dominant element in the system of language. In order to stay alive, the victim supplies any kind of information, regardless of its accuracy. Anything is worth stopping pain. I will give anything, I will say everything so that they do not hit me. I will stop making sense so that they stop the pain.

Thus far, I have attempted to show the limitations of the structural method in the so-called Human Sciences. In the specific consideration of the phenomenon of torture, though, I have maintained one of its basic tenets: the notion that rationality governs the operation of the system. To say that torture "is designed to produce" anything (even mistakes) amounts in the end to embracing the structuralist model of efficiency, of self-sufficiency, of rationality. The ethical risks of such a methodological stand were perceived, as I pointed out, by the members of the CONADEP, who feared that rationalization might conceal a subtle form of extenuation. I also pointed out that many torturers seem to believe that what they do is a legitimate, albeit often unreliable, means of obtaining information, and their conviction would militate even further against the idea that torture is exclusively designed to produce error. It is incontrovertible that it also produces and utilizes truth. A more qualified hypothesis would maintain, then, that torture blurs the line between truth and falsehood. It makes them equivalent, and equally useful from a political point of view: truth is used to suppress specific enemies, blunders to spread intimidation among the general population. And yet, its violence creates something else, something unintended: it inspires in the direct victims and in the general population the courage to face it and to denounce it. In this sense, physical pain does not always work as an efficient structure, as a body.[3] It also errs. It does not always succeed. It is also, and in the double sense, *counter-productive*.

Language, the body, truth: the same figures recur in the testimonies from the torture chambers of the Third World and in the phraseology about the body and language which is heard in the dominant countries, and so often do they recur that one could be tempted to believe in a postmodern conspiracy, which from the shadows, perfidiously, suggested to the automata of underdevelopment the roles in a new theater of cruelty. Or else that torture, like a clown imitating a millionaire sitting in the audience, would take pleasure in enacting a mockery of power. Obviously, neither of those hypotheses is adequate. They are themselves inconclusive caricatures, groundless simplifications. The North, evidently, does conspire, and the South, of course, does tend to imitate. But it is not exactly the protagonists of this drama that conspire or imitate. As a matter of fact, in the United States as well as in Europe, opposition to totalitarian regimes in the Third World is recruited to a great extent among the same sector of the population that "consumes" the postmodern trend. Moreover, the formulations of postmodernism are ignored probably by those who plan and certainly those who execute torture in the Third World.

Language, the body, the absence of truth; if not as a conspiracy or as a mimicry, how to explain then that strange coincidence of concepts and effects between a kind of practice and something like a theory that appear at the opposite extremes of the international political scene? I would like to risk the hypothesis that it is simply that: a coincidence.

But was it really worthwhile to follow the laborious thread of this argument to see it wreck at the end in the almost ludicrous hypothesis of a simple coincidence? Only if we stop for a moment to refine the terms of our study. I am postulating that there is a certain affinity between the experience of torture and the figures of postmodernism. Nevertheless, I am not postulating that such affinity implies a causal or exclusive link between the two. Torture is only one of many ways of besetting the postmodern body with the absence of truth. What makes it special, I think, is that it is prevalent in totalitarian regimes, in dependent economies, in societies, in other words, symmetrically opposed to the ones (France and the United States, for example) where postmodernism has found its readiest followers. Torture seems to have been doing in the South what was being thought in the North, as if, by the principle of the international division of labor, the Northern mind could interpret what the Southern body suffers.

If this conclusion is to be consistent with the cognitive demand for a specific interpretation of torture, which sees the Third World as a peripheral pawn in the international system of power, we must make sure once again that we do not confuse this solidarity, this affinity between the two practices with a causal link. We can hardly imagine a conspiracy of postmodern thinkers who, in the glass and steel darkness of the imperialistic metropolis, would have contrived the distant nightmare of the South. Postmodernism and political terror do not exist as *organs* with a univocal, predetermined function in a centralized *political structure* or *body politic*. They coexist, instead, as *members* of a *political arrangement*. As members, they are in principle politically autonomous and anatomically expendable, supplementary, nonessential. They can only exist in a configuration of power which is no longer exclusively monarchical or disciplinary, but transnational, *corporate, networked*.

In his *La Machine de vision*, Virilio analyzes a rather unusual film by the German director Michel Klier, *Der Riese* (The Giant), randomly assembled with tapes from video recorders installed at crowded points: airports, malls, highways. The result is a feverish sequence of scenes suggesting to the viewer the possibility that her own existence can become at any time the target of this depersonalized, incorporeal *viewing machine,* a state-of-the-art gadget that has finally reverted the traditional terms of the circuit of visual communication: the spectator is being watched, the machine views.

Virilio contends that the functioning of the viewing machine can be used as an emblem of contemporary society. According to him, television is progressively becoming a viewing machine, an empty and panoptical medium, something like an eye without a body, which penetrates the privacy of each home and manages to subject each individual to the tyranny of anonymity.

The preface to *Nunca Más* explained the "spectacle" given by Argentina to "the eyes of the civilized world" by resorting to theatrical metaphors. To express the mechanisms for transmitting images across countries, its authors, educated in the humanist tradition, think more spontaneously about theater, a highly personalized, "cultured" language, than about television, an anonymous mass medium. According to them, the tragedy of Argentina did not reside in being truly barbarous, but rather in *projecting an image* of barbarism, and in becoming a victim of that very image. To translate it into the language of the mass media: political terror was the result of an image of itself over which Argentina had no control, of a kind of "broadcast" or "telecast" that, while local in its origin, still escaped local control.

It is in this sense that I have chosen the expression "networked power" to describe the dispensation of power which is prevalent in the world today. I do not maintain, of course, that television rules the world,[4] but only that the *logic* of television has the same figures as the logic of power relations, as Serge Gruzinki has pointed out in his *La Guerre des images.* Such logic sanctions the autonomy of its members, who can broadcast and connect according to the needs of their own programming. At the same time, the same logic determines their subjection to the same normalizing, empty image. The process of reception becomes progressively chained to the mechanism of emission as the television set turns progressively and insensibly into a viewing machine. Thus, autonomy and subjection are no longer mutually exclusive principles in this *network of power,* typical of advanced imperialism. Indeed, they complement each other in a way that would have been unthinkable for any other previous dispensation of power.

Torture, as it is practiced today, has probably maintained the methods of premodern torture, but has altered its relationship with the state and the individual radically. The archaic, the obscure edifice of human pain has been remodelled, like one of those churches that the crisis of faith and the rise of speculation has turned into fashionable discotheques. Next door, the postmodern trend looks like one of those buildings that multiply historical references, like a discotheque imitating the design of an old cathedral. Both the remodeling and the replica coincide on the landscape of the imperialist metropolis, and their images hobnob in a dialogue of stone and steel, in the same way that the ceremonies of pain and the games of thought get confused in the mind and across the continent.

The recognition of that confusion, of that coincidence, of that connection, does not expose the guilt, or even the responsibility, of the postmodern mind; it merely points to its limits. To venture beyond those limits (to stop before torture), does not mean to take refuge once again in the old myths of subjectivity and truth. It means, on the contrary, to subject those myths to a more ruthless theoretical attack. It also means, it means perhaps above all, that the most ruthless attack has to be more than just theoretical.

NOTES

1. That is why modern drama favors melodramatic play to the detriment of *coup de théâtre*. Modern "man" is a typically Pirandellian hero, a character in a purely discursive tragedy. Who is watching me? Whom am I watching?

2. The word "trend" here is an attempt to translate the Spanish "*onda*," literally "wave," which is used colloquially in Argentina to indicate a style, a mood, or an ideology, in a very indeterminate sense. So, people talk of a *onda punk*, an *onda nostalgica*, or *onda fascista*. The English "wave" in "New Wave" would be closely related. What I am discussing in this paper would be then *la onda postmoderna*.

3. Playing around with Artaud's expression, we could argue that it works like "an organ without a body," insofar as it has a function but not a rationality.

4. During the recent Iraqi war the political power of the television became as literally evident as ever. It also became evident, however, that the power of television is not exactly the power to convey a meaningful image, but rather to create an empty vision. Everybody *watched* the war, nobody *saw* it.

WORKS CITED

Alberdi, Juan Bautista. *La Barbarie histórica de Sarmiento.* Buenos Aires: Pampa y Cielo, 1964.

Argentine National Commission on the Disappeared. *Nunca Más: the Report of the Argentine National Commission on the Disappeared.* New York: Farrar, 1986.

Clastres, Pierre. *La Société contre l'état. Recherches d'Anthropologie politique.* Paris: Minuit, 1974.

Deleuze, Gilles, and Félix Guattari. *Capitalisme et schizophrénie. L'Anti-Oedipe.* Paris: Minuit, 1972.

Deleuze, Gilles, and Claire Parnet. *Dialogues.* Paris: Flammarion, 1977.

Foucault, Michel. *Surveiller et punir.* Paris: Gallimard, 1975.

Gruzinski, Serge. *La Guerre des images: de Christophe Colomb à Blade Runner (1492–2019).* Paris: Fayard, 1990.

Orphée Elvira. *La Ultima conquista del Angel.* Caracas: Monte Avila, 1977.

Plato. "Cratylus." *The Collected Dialogues of Plato*. Eds. Edith Hamilton and Huntington Cairns. New York: Pantheon, 1961.

Sarmiento, Domingo F. *Facundo, o civilización y barbarie*. Caracas: Ayacucho, 1977.

de Saussure, Ferdinand. *Cours de linguistique générale*. Paris: Payot, 1972.

Scarry, Elaine. *The Body in Pain: the Making and Unmaking of the World*. New York: Oxford University Press, 1985.

Virilio, Paul. *La Machine de vision*. Paris: Galilée, 1988.

———. *Vitesse et politique*. Paris: Galilée, 1977.

FRANCES E. MASCIA-LEES
PATRICIA SHARPE

8

The Marked and the Un(re)Marked: Tattoo and Gender in Theory and Narrative

INTRODUCTION: TATTOO AS METAPHOR

Our working title for this paper was "Why the body?", since our goal was to figure out what underlies the intense fascination with the body in current theory.[1] The focus of this volume is only the most recent example of a trend evident in a proliferation of books: *The Body in Pain, Customizing the Body, The Woman in the Body, The Word Made Flesh, Fragments for a History of the Human Body, The Tremulous Private Body,* and two recent special issues of *Representations.*

The physical body symbolically reproduces the anxiety of the social body, according to Mary Douglas. But what of theories of the physical body? What are they symptomatic of in contemporary culture? This is difficult to discern because theory purports to be diagnosis, not ailing patient. But cultural malaise is contagious, and not even the antiseptic practitioner is disembodied and invulnerable in fact. Where can we read the message behind theory's indecipherable prescriptions? Typically we would turn to the body itself since it not only registers but also serves as allegory for culture. Elaine Scarry attributes this tendency to the sheer factualness of the human body. The body is borrowed to lend an aura of realness and certainty to cultural constructs when there is a crisis of belief in society (14), for as Scarry further notes, "given any two phenomena, the one that is more visible will receive more attention" (12). But where do we turn when the body, the very place where social anxiety is traditionally concretized, has been abstracted into theory? Where can we find the living, breathing, secreting, sensing, reacting, weeping body on which today's concerns can be read?

In thinking about Foucault, whose influence underlies much of current work on the body, we were struck by how his conception of history as inscription on the body finds a fuller and more literal treatment in Kafka's "In the Penal Colony." Foucault describes the body as "the inscribed surface of events," and the job of the genealogist as exposing "a body totally imprinted by history" (1980: 148). In Kafka's penal colony this is effected by a torture machine: the ideals of the culture are concretized in a "sentence" pierced onto the flesh of individuals with needles. The text of culture which in other forms is difficult to make out and understand becomes clear to the condemned man: he "deciphers it with his wounds" (204). Thus, the materiality of the body makes language—which Kafka the modernist presents as allusive and ambiguous—into something which can be known and felt. Similarly, Kafka's narrative of cultural inscription can be used to help make tangible the consequences and implications of Foucault's theoretical formulation.

Imagining the process of enculturation as a torturous marking of the body assumes a hostility between culture and the body. Even as contemporary discussions of the body call into question the oppositions essential to modern culture—between the natural and the cultural, the feminine and the masculine, the civilized and the primitive, the unmarked and the marked, the signifier and the signified—their use of the metaphor of inscription may simultaneously reinforce the Western notion of the body as unitary and distinct. It perpetuates images of outside and inside, of encroaching environment and bounded selfhood which seem nostalgic at a time when pacemakers, lens implants, gene therapy, and a whole new technology of artificial limbs have eroded those distinctions in fact.

Although styles of tattoo vary, Western tattoos, in particular, literalize this vision of the body as a surface or ground onto which patterns of significance can be inscribed. They thus serve as a useful metaphor in our exploration of theories that posit culture as "writing on the body." Indeed the relationship between Western tattoo and the body, most commonly one of self-contained images which relate to the body as imposition, may contribute to Western imagery of culture as something imposed from outside. Judith Butler has identified within Foucault's metaphor of inscription on the body the persistence of an assumption that "there must be a body prior to that inscription, stable and self-identical" (130). Believing this inevitably entails the corollary that cultural inscription must do away with the natural body: "In a sense for Foucault . . . cultural values emerge as the result of an inscription on the body, understood as a medium, indeed, a blank page; in order for this inscription to signify, however, that medium must be destroyed" (Butler 1990: 130).

While most Western tattoo similarly treats the skin as undifferentiated surface onto which a predetermined pattern is imposed, some patterns are designed to be applied to specific sites on the body—customarily over the heart, on the upper arm, or on the back. Some styles of non-Western tattoo articulate an even more complex relationship between bodily form and surface pattern. This more abstract tattoo follows and highlights aspects of bodily form. Such abstract patterning which follows the contours of the body— like that on the shoulders of the Igorot man whose photograph James Clifford uses as symbol for the impenetrability of the non-Western (164)—suggests the ways in which a culture constructs the body, reading some features as of significance and failing to recognize others. This then outlines and serves as an image for an everyday cultural practice: the categorization of some body features as, say, sexual. Naming of features such as breasts as sexual parts, for example, helps to define and restrict the erogenous body in Western culture (Butler 114). Still other forms of tattoo, found occasionally in the West and more commonly elsewhere, make use of the features of the body, but transform them into something else: a sailor in "South Pacific" gyrates his stomach muscles to suggest the rolling of the sea under the three masted schooner spread over his chest.

This type of tattoo seems particularly characteristic of the Japanese. In Tanizaki Junishiro's story "The Tattooer," for example, a geisha is transformed by a young tattoo artist obsessed with "finding the perfect woman as a canvas, as it were, for his art" (48). He does not see her as merely surface, however, but uses the form of her body to contribute to his illusion: he drugs her and covers her back with the imprint of a great black spider which, as she awakens, stirs its legs with each of her shuddering, horrified breaths. To the male artist her tattoo is not a mutilation, but a concretization of the hidden message of her body: for him it reveals her true nature as evil and destructive to men. But the woman experiences herself as horribly transfigured. In its power to convert, denaturalize, and defamiliarize the body, this form of tattoo evokes horror at the body's instability, its ultimate inability to serve as a concrete ground of identity.

Here again as with Kafka, narrative illuminates emotions, obsessions, and power hierarchies buried in theory and social practice. Yet Tanizaki Junishiro's story, unlike Kafka's, emphasizes the differential meaning of tattoo for men and women. Kafka's condemned man is used generically as modern human subject, victimized by the caprices of culture and the ambiguity of language. In the world of the penal colony there are only men to stand for the human, and the relationship of all of them to the process of inscription is the same: they are interchangeable. By the end of the story the officer who has manipulated the machine to enact the torture becomes himself the one written upon.

This dramatizes a truism of structuralist theory, that the masculine "corresponds to a non-differentiation between the sexes, to a kind of abstract generality . . . as opposed to which, the feminine is well marked" (Barthes 77). Grammatically, perhaps, but what if we turn to the factualness of the body as allegory? From a woman's common sense perspective, it is the male who appears "well marked." But that is only within the dominant discourse of the West which constructs gender difference as a privative opposition, one of presence and absence, of phallus and castrated. Gender anatomy is better understood as an equipollent opposition, one in which "the two terms are equivalent . . . they cannot be considered as the negation and affirmation of a peculiarity: in *foot/feet*, there is neither mark nor absence of mark" (Barthes 78).

Barthes suggests that linguistically we prefer to understand gender opposition in terms of absence and presence because it permits us to generate a system of parallel pairs like count/countess or poet/poetess which an equipollent opposition like stallion/mare does not. Similarly, at the level of theory, conceiving gender difference as presence or absence of the phallus facilitates the construction of a system of analogous opposition in which the female is associated with nature, the primitive, and the body, while culture, civilization and the mind are seen as male. The body's details, however, can be used to disrupt this dualism. Some feminists, for example, assert the clitoris as talisman to argue that female anatomy is not simply form-fitting sheath.

Our feminist strategy is similar: to turn to narratives about the marked body which, like Tanizaki Junishiro's, highlight gender to show how women's bodies continue to be treated differently, as more natural, and thus to disrupt contemporary theory which frequently envisions writing on the body in ungendered terms, constructing the body as monolithic and undivided, that is, as phallic. In these narratives which foreground gender, we hope to discover what is occluded in much contemporary theorizing of the body. We turn our attention first to a popular film about tattoo contemporaneous with the importation into American culture of Foucault's metaphor of inscription, with the explosion of challenges to traditional definitions of the body, and with feminist assaults on traditional courtship and marriage and the backlash they evoked. To test the historic specificity of the imagistic thinking we discern in the film and theory of the 1970s, we then turn to similar questioning of the relationship between gender and body marking in a narrative written in the mid-nineteenth century by Nathaniel Hawthorne which has become a focus of feminist analysis (see, for example, Fetterly and Johnson). At that time, too, scientific advances were challenging earlier images of the body, theory was attempting to articulate the relationship of man to the bodily and natural, and gender roles were undergoing profound change.

Our goal is to discover how narratives can illuminate theory, and more particularly what narratives about the marked body reveal about the emotional anxieties underlying current theory like Foucault's and its political implication. What crisis of belief might such theories reflect? What agendas do they serve?

INSCRIBING THE MARK: *TATTOO*

Bob Brooks's 1981 film blurs the genres of soft pornography, horror film, and love story; it was marketed to a popular audience and starred the respectable actors Bruce Dern and Maude Adams. It tells the story of a tattoo artist (Dern) hired by an advertiser to paint fake tattoos onto swimsuit models and who becomes obsessed with one of these women. He seeks to establish an old-fashioned relationship with her, one where he orders dinner for her, brings her flowers, pays the tab, and protects her. He drives her home when she says she is afraid of being raped, and says "Love should be like a shield. If you love someone, you should protect them."

"Most women can protect themselves these days," she replies. The contradiction between this statement and her original request that he accompany her home because she was scared suggests female duplicity. Was she merely flirting, playing to his manhood, or manipulating him out of desire in her request? Is she now hiding her vulnerability to preserve her autonomy? Or is she simply contradictory, flighty, irrepressible, a confirmation of Dern's suspicion that women's word cannot be trusted? In all events, her reply calls into question his traditional notion of male chivalry and can be read as a challenge to his conviction that women need protection, a challenge we are led to suspect will be proved hideously naive as the story unfolds. With sinister foreshadowing the film dwells on tattooing as a form of protection: when a male customer seeks out the tattoo artist to get a "mean tattoo" which will protect him against thugs, Dern confirms the validity of his request, saying "Tattoos have been used for centuries to ward off evil spirits."

The story turns on two events which, when read together, expose men's anxiety over a "liberated" female sexuality which they fear cannot be trusted. When Adams uses the word "fuck" casually in conversation, Dern flies into a rage saying: "Things must last. Fucking is only instant love, it doesn't last." He seems horrified by this suggestion of her casual attitude toward sex. Although the sexual freedom of the 1970s has been understood by some feminists as the ideological product of a misogynist culture allowing men increased access to women's bodies, it is also true that women did become more sexually active outside of marriage and more open in discussing their sexuality. During that decade a range of behaviors which were troublesome to society—from male sexual impotence to divorce—was blamed on women's failure to perform their traditional gatekeeper role. Dern's response reflects the fear which men seem to have experienced of an unbounded female sexuality out of their control.

The second event plays on male fear of female perfidy: after Adams becomes suspicious of Dern's excessive interest in her, she attempts to avoid him, having her secretary tell him that she is out of town. Unbelieving, Dern watches her apartment, sees her enter with her boyfriend, and catches her in the lie that he has suspected. This is part of a repeated pattern of voyeurism in the film beginning with the camera eye's focus on Adams modeling. It continues as Dern slowly scrutinizes her photograph when placing it under a transparent overlay on which he practices drawing the tattoo images he has been hired to paint on her body. The association of the gaze with male desire and the fear of loss of control, and with attempts to preserve distance is underscored when Dern goes directly from rejecting uncommitted intimacy with Adams to finding sexual pleasure at a peep show. This is later echoed in his behavior with Adams: until he has assured himself of her total submission, he refuses intercourse with her, demanding instead that she strip for him and masturbate while he observes through a hole in the door.

In many ways the movie sets up Dern's efforts to control Adams in opposition to her freedom as a model and pathologizes Dern's voyeurism; it downplays the fact that as model Adams is also subject to the gaze. A model in our society must subject her body to extreme disciplinary regimes in order to bring it into conformity with a cultural ideal which she has internalized: "this self-surveillance is a form of obedience to patriarchy. It is also the reflection in woman's consciousness of the fact that *she* is under surveillance in ways that *he* is not, that whatever else she may become, she is importantly a body" (Bartky 81, emphasis in original).

Many women participate willingly in this system because it holds out for us an illusion of potential freedom and power: the perfect body will be for a woman the ultimate currency of exchange. It will allow her to escape the trap of a sexually exclusive relationship or of having to depend on a man financially. This ideology of freedom was prevalent in the 1970s, encouraging middle-class women to see joining the work force as a form of liberation. While granting some women increased autonomy, even as it compounded the double burden for many others, this shift in ideology from women as homemakers to women as autonomous workers has acted to obscure capitalism's failure in the 1970s to provide the affluence to families with only one salary that was available in the 1950s.

The movie dramatizes the conflict between this idea of sexual and economic freedom of choice for women and Dern's nostalgic endorsement of a more traditional ideal of dependency, protection, purity, and fidelity. Rather than valorize this sentimental attitude, the film presents Dern as pathological and terrifying, to the female viewer at least, thus reinforcing our commitment to an ideal of autonomy which may serve society's current needs.[2] The extremity of Dern's obsession is revealed when he drugs and kidnaps Adams, takes her to a deserted and isolated house, and begins a process intended to enforce her faithfulness, inscribing on her the codes of this older system of values: he tattoos her chest, shoulders, and back.

Like the geisha in Tanizaki Junishiro's story, Adams is horrified, experiencing the tattoo as worse than death. After she awakens and realizes what Dern has done to her, in agony and rage she breaks the mirror that reflects her deformed image. For both women, having their bodies written by male culture is a form of mutilation. But for Dern, the tattoo is a mark of his commitment and, thus, a protective shield for the woman on whom he places it. Dominated himself by a powerful father, Dern feels oppressed by his cultural responsibilities and resentful that women elude culture's mark: he seeks to restrict her, as he feels he has been restricted, by literally writing on her body. In Kafka's "In the Penal Colony" it is prisoners who have failed to uphold the law who must be imprinted with it. This film *Tattoo* suggests that those who fail to uphold the law are likely to be women. As Freud would have it, our superegos are deficient. We cannot be trusted to ascribe to "the right" (male) cultural values. The film, thus, taps into a central anxiety of Western culture: males' fear that women will be "disloyal to civilization," failing to uphold the sex/gender system which sustains male power. Perhaps because women are less invested in this system due to the very marginalization of women on which the system depends, we are open to suspicion of betrayal and treachery and the conviction that we must therefore be controlled.

It is, of course, significant that in Western culture men choose to have their bodies marked more frequently than women. Clinton Sanders's study of prisoners suggests that they see their tattoos as marks of identity which cannot be stripped from them as their clothes and other personal possessions have been. These tattoos, thus, speak a defiance against an oppressive authority (40). Study of tattooed individuals in the larger population confirms the notion that tattoos are primarily sought by men as a mark of identity, symbolizing personal interests, associations, and most generally, a rebellious masculinity. They serve "to separate the [male] individual from the normative constraints of conventional society" (Sanders 50). Dern's character in the film underscores the notion of tattoos as a form of self-symbolization, as a mark of identity. Applied to a woman, they indicate not her identity but that of the man to whom she is tied; they signal sexual exclusivity, fidelity to one man, hence her identity as someone's exclusive property. When asked by Adams what kind of women get tattooed, Dern replies "women who need the mark in order to exist, women who are ready to be identified, ready to belong," to belong, that is, to a man.

Yet the restrictions imposed by the tattoo constrain more than Adams's sexual freedom. Once tattooed by Dern, Adams will no longer be able to work as a fashion model adjusting her body and look to the demands of various shoots and photographers. She can no longer use her body as a medium of exchange; its meaning and use have been severely restricted. Dern strives to give her body a univocal reading, to allow it to speak only his meaning, faithfulness to him. Tormented by female autonomy and instability, fearful of her capacity to manipulate language and he suspects to lie, Dern turns to the woman's body as a site to control the play of the signifier, to dictate who can read it and what it can say.

Unlike the men in Western culture who freely choose or refuse to be tattooed, thus choose or refuse symbolic marking, Adams's character as model, like women in general, cannot escape it. Dern's mark simply exemplifies a cultural process by which her status as female is constructed and controlled. Monique Wittig elaborates on the implications of structuralist insights into the functioning of the mark of gender. Gender designations in such languages as French and English which particularize the female allowing the male speaker to appropriate the universal for himself buttress and reinforce power hierarchies:

> Sex under the name of gender, permeates the whole body of language and forces every locutor, if she belongs to the oppressed sex, to proclaim it in her speech, that is, to appear in language under her proper physical form and not under the abstract form, which every male locutor has the unquestioned right to use. (Wittig 66)

It is this marking, this particularization, that denies the female subjectivity since "a relative subject is inconceivable" (Wittig 66). Thus, the mark of gender in language and society precludes women from taking up the position of Absolute Subject, even as it constructs us as female. The women in the film who come to Dern to be tattooed in order to exist at all epitomize women's position in Western culture: we are constituted relationally in language. It is only by being marked in relationship to the unmarked male that women come into being, a being that is inevitably partial: it is the being "woman," not the "I." Similarly, the wedding ring, the woman's married name, or Dern's tattoo grant identity and confer status to the woman through signifying her relationship to an individual man.

Dern's character displays the anxiety in the 1970s that these traditional symbols could no longer be trusted as signs of permanent commitment. Like the tattooed images that he painted on Adams's body earlier in the film and that he subsequently watches being washed away, they are ephemeral, he fears, empty symbols that have lost their authenticity. This association of actual tattoo, as opposed to painted imitation, with the authentic is underscored by allusions to Japanese ritual in the film.

This use of the East in a naive ideology of authenticity is part of the same tradition in the West which has constructed "the Orient" as the exotic and "Oriental" women as highly erotic in their sub-missiveness. Thus, the use of the non-West to indicate the genuine and the real parallels and reinforces Dern's nostalgia for a sexual intimacy which guarantees true commitment and a certainty about the woman's fidelity. These allusions to the East underscore Dern's longing for a woman who is thoroughly sexualized but exclusive.

The film, however, inevitably fails to assuage male anxiety. After his work on her is complete, Dern exposes his tattooed body to Adams and has sexual intercourse with her. But his mark has failed to ensure her trustworthiness: as he begins to reach orgasm, she stabs him in the back with one of the tattooing instruments that he has carelessly left within her reach. Thus, the film enacts a male fear repeatedly voiced in Western culture: that loss of control in orgasm can result in male death. The female fear which the film arouses about male control is not assuaged either despite Adams's seeming triumph. Even after his death, the tattoo Dern has placed on Adams's body ensures that she cannot have a relationship with any other man without revealing Dern's mark and hence that she has a past, that she can never be exclusively his.

Adams's character has known the relative freedom to work and to choose sexual partners, but Dern's mark will curtail her model-ing. It cuts her off from her milieu and acts as a permanent reminder of her subjection. She experiences it not as affirming of identity but as a horrific restriction. In setting up this contrast between women's traditional entrapment in monogamous relationships and their comparative freedom in the 1970s, the movie obscures the ways in which "liberated" women continue to be marked as feminine. Understanding gender differences as malleable, many feminists in the 1970s valorized androgyny, calling for a blending of masculine and feminine traits in individuals of both sexes in order to free both women and men from the constraints of gender. Recent feminist theorizing (see, for example, Haraway 1988) has urged, instead, that males call attention to their gender, that they learn to mark the male condition, to acknowledge their own relationship to the human as partial.

Theory since structuralism, even that sympathetic with these more recent feminist insights, risks falling into the same trap as celebrators of androgyny, assuming that since the sign is arbitrary in linguistic theory, it can also be so in practice. But the connotative meaning of an individual sign cannot simply be changed and stripped of the socio-historic conditions associated with it. To individuals, maleness and femaleness are not arbitrary signs, but historically conditioned givens which cannot simply be wiped away, whether through the construction of alternative cultural myths or sex change operations. Those postmodern theorists who ignore gender, who ignore the difference that difference makes, also erase the female, even as the rediscovery of the body might help them break with a conception of the subject as unmarked, abstract, and universal. They thus perpetuate a privileging of the male and risk reinscribing the very monolithic subject that they have attempted to deconstruct.

EXCISING THE MARK: "THE BIRTHMARK"

How does it happen that attempts to revise symbolism in order to expose and change the power dynamics it represents and reinforces can so backfire? The resistance arises both in society, in people threatened by a rethinking of structures which make them comfortable, and in the functioning of symbols themselves. Because they have an arbitrariness that the social relations they emblematize do not, symbols can be radically reconfigured only to reinforce the status quo in new ways. This is evident in a system which constructs difference in terms of privative oppositions represented by marked and unmarked terms. Barthes notes that "some linguists have identified the mark with the exceptional and have invoked a feeling for the normal in order to decide which terms are *unmarked*" (76). Working from content, from the signified, these linguists then argue that the marked condition can be signaled either by the addition or subtraction of a significant element. The presence of the penis can be constructed as the norm and the feminine condition marked by its absence—a conceptual framework evident in the equation of woman with the castrated. As elaborations of symbolic thinking, narratives too exploit the arbitrariness of the sign; a contrast of the film *Tattoo* with Nathaniel Hawthorne's story "The Birthmark" demonstrates how a male intent on wiping away the mark from the female may be just as destructive as the male intent on placing his mark on her.

The story describes a male scientist at the end of the eighteenth century who becomes obsessed with removing a birthmark from his wife's cheek. Before their wedding he had not even noticed the mark which "bore not a little similarity to the human hand, though of the smallest pygmy size" (149). Yet Hawthorne stresses that it was one of her attractions for other men: "Georgiana's lovers were wont to say that some fairy at her birth hour had laid her tiny hand upon the infant's cheek, and left this impress there in token of the magic endowments that were to give her such sway over all hearts" (149). Women, by contrast, found it hideous, calling it "the bloody hand" (149). Thus both read it as feminine, women associating it with menstruation and parturition, men with beauty.

Social historians record the impact on gender ideology of profound economic and social changes in America between 1825 and 1850. "The emerging ideology of individualism," Leverenz notes, "erected an ideal of free, forceful and resourceful white men on the presumption of depersonalized servitude from several subordinated groups" (39). An accompanying cult of spirituality and domesticity, of fulfillment through tender self-sacrifice, was emerging for white women. The hint of sensuality and physicality suggested by Georgiana's birthmark is intolerable in a wife; it becomes the quest of Aylmer, her scientist husband, to enforce her spirituality, directing all his scientific energies toward excising the birthmark: "what will be my triumph," he declares, "when I shall have corrected what nature left imperfect in her fairest work!" (152).

In his own life, Hawthorne felt similarly compelled to enforce a rigid ideal of femininity on his daughter Una. His impression of her as a child stresses the confounding of gender categories: "the child appears to him an anomaly, neither male nor female and yet both. She strikes him as not human, in uncanny moments, because she does not conform to the definitions that organize his perceptions of the human" (Herbert 285). According to Herbert, this was profoundly disturbing because "the gender system that ascribed nurturant tenderness to women and combative individuality to males was conventionally regarded in Hawthorne's time as a law of nature and of nature's God, a matter of universal essences that were at once biological and ethereal" (287). Herbert records Hawthorne's ambivalence toward this gender system and the anxieties it roused in him since "he felt his own character to be anomalous in relation to the prevailing standard of masterful public manhood" (285). This led him to interrogate the natural foundation of manhood and womanhood in sympathy with feminists of his day, even as he lashed out at those women who displayed a forceful manliness which he feared he lacked (Herbert 285–286). Hawthorne's ambivalence toward activity and assertiveness in women affected his daughter: "It is as though she had been covertly encouraged to repudiate, by some communication of Hawthorne's own ambivalence, the very model of womanly fulfillment she had been enjoined to follow" (Herbert 292).

Hawthorne similarly encodes a double message in "The Birthmark": he asserts nature's awesome power to control human existence even as he calls into question the prevailing view that nature dictates rigid gender roles. His narrative of Aylmer's increasingly desperate and dire efforts to remove the birthmark—first with chemicals, then when they fail with surgery—in order to make his wife totally pure, perfect, and beautiful suggests the enforcement of spirituality and passivity in women which Hawthorne saw in his period and felt compelled to carry out on his daughter.

Through his narrative of Aylmer's disastrous quest, Hawthorne reveals the consequences of his society's rigid construction of femininity: Georgiana can become "the perfect woman" only when all traces of what was not the dominant image of ideal womanhood—whether physicality, sin, masculine self-assurance, or desire—are erased, that is in death. Georgiana's birthmark, then, stands for the disruptive elements that drove so many nineteenth-century women, as well as their fictional representations, to frustratingly contradictory aspirations and to madness. Hawthorne underscores this when Georgiana poignantly asks Aylmer "Cannot you remove this little, little mark, which I cover with the tip of two small fingers? Is this beyond your power, for the sake of your own peace, and to save your poor wife from madness?" (152).

Aylmer's ministrations to his wife can be read as a horrified criticism of the sort of scientific treatments—enforced restraint, rest cures, and medical electricity—to which Hawthorne's daughter Una was subjected when her ambivalence expressed itself in the neurasthenia, mental disturbances, and hysterical violence prevalent in the period. The whole effect of the story is to make this enforcement of a cultural and aesthetic ideal of femininity seem excessive and unnecessary, indeed a challenge to nature. Yet this is paradoxical: the story affirms nature's power to defy man's meddling by having Georgiana die, even as it calls into question the prevalent creed of gender's biological determinism, its natural construction. Georgiana's birthmark is finally read as emblem of a stubbornly unchangeable natural humanity: "It was the fatal flaw of humanity which Nature, in one shape or another, stamps ineffaceably on all her productions" (149). Hawthorne uses Aylmer, the white male scientist, as representative of culture while his wife bears the mark of nature, only finally to expose and disrupt that pairing of binary oppositions by suggesting that all human creatures are similarly marked, hence that Aylmer must be as well. His quest entails projecting his own imperfection and mortality onto his wife while denying it in himself, then seeking to eradicate it altogether.

Thus Hawthorne appears to have negotiated his own gender ambivalence through constructing narratives in a high art tradition that valorized ambiguity, multiplicity, and paradox. In them he could assuage the anxiety provoked by proposing that gender is not immutably inscribed in nature by asserting that nevertheless nature is powerful and provides a much needed brake on nineteenth-century idealizations of science, progress, and perfectibility. In arguing for women as human, Hawthorne challenged pressures in his culture to make women unitary, to make them signify only femininity, even as he reinforced an ideology of the natural.

The pressures to redefine manhood and its relationship to all that it is not, whether the feminine or the natural, whose emotional consequences are explored in Hawthorne's story also find theoretical exploration in writing of the period. Like "The Birthmark," Emerson's essay "Nature" is a critique of late eighteenth-century science. In it Emerson advocates a revision of the dualistic thinking that underlies the conceptual opposition of nature and culture. He challenged a rationalistic science which conceived nature simply as matter from which the mind and spirit were divorced. He seems almost to anticipate Douglas and Scarry in his argument that human beings see in the tangible body and in nature metaphors of mental and spiritual concerns: "Whilst the abstract question occupies your intellect, nature brings it in the concrete," he declares (42). Throughout the essay, Emerson makes the point with messianic enthusiasm:

The use of natural history is to give us aid in supernatural history; the use of the outer creation, to give us language for the beings and changes of the inward creation. Every word which is used to express a moral or intellectual fact, if traced to its root, is found to be borrowed from some material appearance. (14)

At times Emerson even seems to acknowledge, as Douglas and Scarry would, that this is an effect of our capacities for processing information and of a tendency to project our needs and emotions onto nature, to construct it through our conceptual frameworks. Commenting on his pleasure in seeing boughs tossing in a storm he acknowledges "the power to produce this delight does not lie in nature, but in man." Then he goes on to add more characteristically, "or in harmony of both" (7). For most often Emerson presents nature as a sort of mood ring, responsive to man as, for example, when he declares "nature always wears the color of the spirit," or in his assertion that "every natural fact is a symbol of some spiritual fact" (15); that "the world is emblematic. Parts of speech are metaphors, because the whole of nature is a metaphor of the human mind" (20).

In his conception of nature as enthusiastic and appreciative backdrop for man's exploits, Emerson contrasts with Hawthorne. While Hawthorne used the indominability of nature as check on man's presumptions to mastery, Emerson argues that nature can best be mastered when it is understood as inviting to man and responsive to him:

Nature stretches out her arms to embrace man, only let his thoughts be of equal greatness. Willingly does she follow his steps with the rose and the violet, and bend her lines of grandeur and grace to the decoration of her darling child. Only let his thoughts be of equal scope, and the frame will fit the picture. (12)

Here we see the superimposition of mutually reinforcing binary oppositions, that had been disrupted in Hawthorne, equating civilization with male and picture, while nature is female and frame.

Yet even within Emerson there is slippage; we can only understand how man can be at once the "child" of nature and its dominator when we recognize Emerson's role in articulating the emerging gender ideology about which Hawthorne displayed ambivalence. In Emerson, too, the developing ideology of manhood stirred self-

doubt which he negotiated through writing; yet rather than constructing narratives which explored the emotions and costs of this
ambivalence, Emerson sought "spermatic, prophesying man-making words" (Emerson quoted in Leverenz 39). Emerson constructed
out of his own insecurities a prose which "inspires feeling of access
to manly power" (38), according to Leverenz, who argues that it has
been an important source of Emerson's appeal in recent years, in the
period of post-structuralism and the film *Tattoo*, for critics like
Harold Bloom. "Emerson's manly eloquence really transforms his
sense of his female defects, especially his sense of powerlessness and
his inability to make up his mind" (Leverenz 40, drawing on
Cheyfitz). Thus, Emerson constructed a manhood which inhered in
never being pinned down, constructing an ideal of "imperious nonchalance, a manliness beyond competition" (Leverenz).

Rather than question the ideology of manhood, Emerson
sought to redefine it: "In a society that defined manhood com-
petitively by possessiveness and possessions, Emerson would redefine manhood paradoxically by abandonment and self-possession"
(Leverenz 40). This is then parallel to his call for a new understanding of nature and man's relationship to it, one which would see his
dominance not as opposition and conflict, but rather as desirable
stewardship. Yet as Leverenz emphasizes,

> Though Emerson challenges the social definitions of manhood
> and power, he does not question the more fundamental code
> that binds them together. . . Emerson's ideal of manly self-
> empowerment reduces womanhood to spiritual nurturance
> while erasing female subjectivity. (39)

Thus Emerson's image of mental health, whose gender assumptions
Leverenz rightly underscores, is "the nonchalance of boys who are
sure of a dinner" (Emerson's *Self-Reliance,* quoted in Leverenz 39).

Emerson's inability or unwillingness to really rethink gender,
then, sets the limits on his challenge to contemporary views of
nature as his assumptions are governed by superimposed binaries.
This is evident in a particularly striking passage in "Nature" where
he offers examples to illustrate how nature adorns any "noble act":

> When the bark of Columbus nears the shore of America; before
> it the beach lined with savages, fleeing out of their huts of
> cane; the sea behind; and the purple mountains of the Indian
> Archipelago around, can we separate the man from the living

picture? Does not the New World clothe his form with her palm-groves and savannahs as fit drapery? (12)

Here another dimension is added: nature is not only associated with the motherly and the nurturant, but with the eagerly awaiting savage who like the sea and purple mountains serves as backdrop for white male exploits. The imagery of clothing further reinforces the opposition between "Columbus" and the "savages," the draped civilized as opposed to the naked primitive. With his phrase "fit drapery," Emerson illustrates the paradox in the way such binaries function, at once constructing the dressed state as normal and given, even as it associates the natural with nakedness.

In Hawthorne, too, the natural is associated with the non-Western, Aylmer's dark servant Aminadab:

This personage had been Aylmer's underworker during his whole scientific career, and was admirably fitted for that office by his great mechanical readiness, and the skill with which, while incapable of comprehending a single principle, he executed all the details of his master's experiments. With his vast strength, his shaggy hair, his smoky aspect, and the indescribable earthiness that encrusted him, he seemed to represent man's physical nature; while Aylmer's slender figure, and pale, intellectual face, were no less apt a type of the spiritual element. (153)

Yet Hawthorne's story ultimately disrupts the power hierarchy of mind/body, science/nature and the final triumph—"the gross fatality of earth over immortal essence"—is heralded by Aminadab's "hoarse, chuckling laugh" (165). Rather than celebrating the white man's conquest like the savages welcoming Columbus, Aminadab revels in the failure of his master's schemes as Hawthorne's story realigns nature with the normal, making Aylmer's ambitions seem excessive and abnormal.

In his image of Columbus, Emerson uses a pictorial metaphor as he had earlier, comparing man to the picture for which nature is frame. Here, with his rhetorical question "can we separate the man from the living picture" (12), he implies the union of figure and ground. It is clear, however, that this is accomplished only by the subsuming of nature, the feminine, and the primitive to the Western male: "an act of truth or heroism seems at once to draw to itself the sky as its temple, the sun as its candle" (12).

Barbara Johnson explores the implications of such imagery and its association with the erasure of the woman in Hawthorne's "The Birthmark," which she reads in conjunction with Gilman's "The Yellow Wallpaper" and Freud's interpretation of the Dora case. She interprets Georgiana's birthmark as an "overinvested *figure* inscribed on a page that should be blank," a sign of woman's multiplicity (258, emphasis in original). Thus, she sees "The Birthmark" as a "story of fatal clitoridectomy," when the clitoris is understood, as it is for Gayatri Spivak and Naomi Schor, "as synecdoche for the possibility that the world could be articulated differently" (Johnson 246).

Oddly, though, Johnson reads Aylmer's efforts to reduce the "woman to 'all,' to ground, to blankness," as an effort "to erase sexual difference" (259). This seems to echo Judith Fetterly's claim that "what repels Aylmer is Georgiana's sexuality; what is imperfect in her is the fact that she is female; and what perfection means is elimination" (25). Yet our reading suggests that the story is less about erasing than enforcing sexual difference. To perform clitoridectomy is to remove one aspect of her multiple sexuality: when the organ that has been read as analogous with the male's is excised, woman becomes exclusively vagina, the inverse of the male and the privative opposition is concretized in bodies. Removing the woman's mark here, like marking her in the film *Tattoo*, is an effort to make her univocal. It is not her sexual difference, or even her sexuality, that is erased, but rather her activity, her self-definition, her supposed masculinity.

Johnson's title "Is Female to Male as Ground Is to Figure?" explicitly echoes that of Sherry Ortner's "Is Female to Male as Nature is to Culture?" suggesting feminism's long struggle to understand the relationship of a system of privative oppositions in language to women's oppression in the world. Ortner stresses the universality and timelessness of a master narrative that, she claims, always sees women as closer to nature and nature as inferior to culture. Johnson, by contrast, emphasizes the bringing to light of lost alternatives, the need to reconstruct gender opposition as a recursive picture—one in which ground can be seen as figure in its own right, analogous to Escher's hands, each drawing the other—or in our terminology, as an equipollent opposition. A simple reversal of binary terms cannot bring about a truly recursive picture, as our analysis of *Tattoo* and "The Birthmark" makes clear. For either the marking or the removal of the mark can serve as a representation of the imposition of a restrictive, unitary femininity. In one instance, Georgiana is marked by nature as multiple while in another the woman in *Tattoo* is seen as multiple in her blankness, as needing to be marked by culture to make her meaning unitary.

Hawthorne anticipates this insight. Coupling "The Birthmark" with his most famous work *The Scarlet Letter*, we can see that Hawthorne understands efforts to inscribe the female body as similar in motive and effect to Aylmer's removal of Georgiana's birthmark. His story of the efforts of a Puritan community to imprint on Hester Prynne an incontrovertible and unambiguous cipher of her sin, adultery, ultimately shows how differences in perspective and historical circumstances destabilize the sign: as Hester lives a life of service to the community, "many people refused to interpret the scarlet A by its original significance. They said that it meant Able" (160). As he did with Georgiana's birthmark, so with Hester's A, Hawthorne records the proliferation of interpretations by rank, gender, and circumstances as well as over time: in the end "the scarlet letter ceased to be a stigma" (261), on Hester's body that is, again suggesting that the female is polysemous. Because the woman is seen as multiple, whether because of her dual sex organs, or her multiple and conflicting roles as Ortner would have it and as the film *Tattoo* suggests, or because in language her femininity and humanity are constructed as at odds, she appears to escape cultural definition. By contrast, the male experiences himself as having no escape. Hawthorne concretizes this difference in the contrast between Hester's embroidered letter and that on the bare breast of her lover Dimmesdale where it is "imprinted in the flesh" (256).

Thus, we are back to an early scene in the film *Tattoo* where men with actual tattoos pose beside female models whose apparent tattoos are merely surface paint. This contrast dramatizes the mythic perception that the cultural codes which have painfully penetrated male flesh lie easily upon the surface of the female body. This is made explicit in Dimmesdale's third person self-accusation to the people of New England from the scaffold:

> He bids you look again at Hester's scarlet letter . . . with all its mysterious horror, it is but the shadow of what he bears on his own breast, and that even this, his own red stigma, is no more than the type of what has seared his inmost heart! (253)

This suggests a pervasive conviction in the metaphoric thought of the West that it is the male who is unitary, and thus, that it is his body which does not lie.

THE BODY AS TRUTH

Indeed, uncovering this myth—that it is in male flesh that one can read the truth of culture—confirms our suspicion that when postmodern theorists claim that *the* body is used as a site where culture's truths and anxieties are concretized, it is the male body being read as generic, just as in Emerson a particular ideology of white Western manhood became the only possible form of "man" to set in relationship to a nature which subsumed the woman and the native. This is nowhere clearer in contemporary theory than in Foucault's *The History of Sexuality* and can easily be demonstrated by contrasting his reading of Diderot's "Les Bijoux Indiscrets" which does not attend to gender differences with Jane Gallop's that does.

Diderot's story tells of a sultan whose magic ring will make a woman's vagina speak. In Foucault, this is simply a metaphor for modern society's paradoxical attitude toward sex: "speaking of it *ad infinitum*, while exploiting it as *the* secret" (35). It represents the contemporary condition in which "each one of us has become a sort of attentive and imprudent sultan with respect to his own sex and that of others" (79). The generic masculine pronoun in this sentence is typical of Foucault's tendency throughout the book to discuss sexuality without differentiating its impact by gender. Foucault identifies the modern sultans who interrogate sexuality as "priests, demographers, psychiatrists, and physicians," but does not acknowledge that these are masculine authorities who have had the power to define the feminine. Gallop, by contrast, makes it clear that the sultan is a male empowered by the magic ring to make his beloved's vagina speak the truth of her feelings for him. Her reading suggests that Diderot's story shows not only a generalized Western preoccupation with the confessional mode, with the desire to have sexuality speak, but also the specifically male wish to arrest the duality and possible double-dealing of the woman.

In the same way that Gallop's "Snatches of Conversation" illuminates the difference that gender makes which is obscured in Foucault's analysis, so Susan Gubar's "'The Blank Page' and the Issues of Female Creativity" can help us rethink the metaphor of cultural inscription on the body as it applies to women. Gubar shows how the predominant Western cultural metaphor of writing on the body is one in which the woman is envisioned as the blank page awaiting inscription and the male as the writer. The dilemma for the woman artist is thus how to reconceive this metaphor to achieve artistic agency. Our reading of cultural myths concretized in popular film and "high" art shows that men as well as women are conceived by male artists as surface. However, despite this apparent similarity, gender dynamics pervade these myths, dictating who can write and the nature of inscription in individual cases.

As was clear in Kafka's penal colony, the male is inscribed by a higher power, by the cultural machine. Similarly Dimmesdale is marked by a higher power, whether it is divine judgment, the magic conjuring of Hester's husband Chillingworth, or the effect of his Puritan religion and the guilt it instills in him eating away at him from inside or impelling him to torture himself. Females in these stories, by contrast, are marked (or unmarked) by the male. The male fears her not because she may inscribe him, but because he thinks she may be freer than he is, that she may escape the mark that fixes meaning, carry multiple meanings, and thus remain ambiguous. In this way, she is seen as outside of culture, and therefore equated with the natural in a binary system where there is no place else to go. But, of course, the female is not actually closer to nature or outside of cultural constraints; feminist studies over the last twenty years have repeatedly revealed the way in which women's bodies serve as a site of masculine control. What then are we to make of stories that insist on women's multiplicity and ambiguousness, our supposed elusiveness and mysteriousness? They not only reflect male anxiety over the chaos men suspect that the feminine can unleash, but also act as cautionary tales: they suggest to men that women's power and bodies must be brutally imprinted to ensure that we will conform to male cultural values.

Gubar offers an image for the kind of female power that might be threatening to men in the story of the blank sheet she adapts from Isak Dinesen. Hung in a gallery of princesses' wedding night sheets where the bloody spots testify to virginity, the one blank sheet has no one meaning:

> Was this anonymous royal princess not a virgin on her wedding night? Did she, perhaps, run away from the marriage bed and thereby retain her virginity intact? Did she, like Scheherazade, spend her time in bed telling stories so as to escape the fate of her predecessors? Or again, maybe the snow white sheet . . . tells the story of a young woman who met up with an impotent husband, or of a woman who learned other erotic arts, or of a woman who consecrated herself to . . . chastity but within marriage. (259)

Similarly, Adams's unmarked body, or Georgiana's hand, yields multiple interpretations which men feel compelled to control.

Gubar's analysis acknowledges that the metaphoric pattern of conceiving women as blank page is myth, even as she shows the tremendous intellectual and imaginative burden it places on women struggling to envision themselves as creative agents. Like Johnson, Gubar affirms alternative stories even as she helps to explain the fear that underlies their repression: if no story is written by men upon the women's body, she may make up her own that threaten male control. While feminists like Johnson, Gallop, and Gubar acknowledge the differential gender impact of these myths, and explore its implications for women, other contemporary thinkers do not. Thus they mimic the male tattooist, eager to impose his condition on the female, to deny her difference, in fact to do away with duality. Their focus on the monolithic body thus echoes the males in these stories, eager to establish a unitary truth.

This contemporary impulse is not merely theoretical; it also has an impact on real bodies. Jean Bethke Elshtain's recent analysis of what she terms the sex-neutrality model, the view that gender is a cultural construct, shows the continued dangerous consequences for women of trying to simply wipe gender difference away. She suggests that both the anorectic and the "Jane Fonda super-fit" body can be understood in relationship to the increasing tendency to evaluate the female body by the terms set by the dominant body-self, "the normative, techno-male of the modern West" (130). Starvation and extreme dietary and exercise regimentations result in a female body that is the literalization of contemporary American society's trope of sex-neutrality and gender equality.

At least since the seventeenth century, "the woman" has been coded as body in Western culture, and the body has been seen as feminine (see, for example, Barker, 1984). What does it mean that it is the male who is now seen as embodied? What does our analysis reveal about the popularity of the body as a site for contemporary theorizing? Reclaiming male embodiment might seem a significant effort toward redefining traditional gender oppositions, and notions of the body as constructed in culture may seem to disrupt standard truths. However, the body seen as cultural construction has become a construction that is itself taken as incontrovertible truth. Bodies are now used paradoxically to deny the natural and to "embody" the contemporary homily that there is nothing outside of culture.

Once everything that had formerly been conceived of as natural is absorbed into the cultural, men can claim it as their own. They can have bodies when the body is dissociated from the natural and, therefore, from the feminine. Thus, like Emerson, some contemporary theorists intent on deconstructing binary oppositions fail to construct a thorough-going critique because of their attitudes toward gender. Rather than displace traditional gender categories, these theories subsume once again the female to the male, obscuring the way in which power relations continue to produce different bodies differently. In this paper we have looked at two periods of profound changes in which gender roles as well as other defining binary oppositions were under stress. Our reading of narratives about the white female body exposes what is missing in theory. While Emerson erases female subjectivity, contemporary theory necessarily erases her body as well. To claim the body as male without a dangerous suggestion of femininity, it is necessary to remove the female altogether.

NOTES

1. We wish to thank Mary Margaret Kellogg for her generous support of the Women's Studies Program at Simon's Rock College of Bard and its Faculty Development Fund. Support from that fund made it possible for us to present an earlier version of this paper at the Annual Meeting of the American Ethnological Society, Atlanta, Georgia, April 1990. We are also grateful to Francis C. Lees for his generous assistance and to Jamie Hutchinson for thoughtful conversations over the years which inspired and challenged our analysis of Hawthorne and Emerson. Nevertheless, he bears no responsibility for the views we express, recalcitrant colleagues that we are.

2. Even as ideology mobilizes women to work to mask economic threats to the familial status quo, it also stirs anxieties to control any disruptive effects of the change. A later film much more sympathetic to male anxiety about female autonomy and the threat it poses to traditional domesticity is *Fatal Attraction*.

WORKS CITED

Barker, Francis. *The Tremulous Private Body: Essays on Subjection*. London and New York: Routledge, Chapman, and Hill, 1985.

Bartky, Sandra Lee. "Foucault, Femininity, and the Modernization of Patriarchal Power." *Feminism and Foucault: Reflections on Resistance.* Eds. Irene Diamond and Lee Quimby. Boston: Northeastern University Press, 1988. 61–86.

Butler, Judith. *Gender Trouble: Feminism and the Subversion of Identity.* New York: Routledge, 1990.

Clifford, James. *The Predicament of Culture: Twentieth Century Ethnography, Literature, and Art.* Boston: Harvard University Press, 1988.

Elshtain, Jean Bethke. "Cultural Conundrums and Gender: America's Present Past." *Cultural Politics in Contemporary America.* Eds. Ian Angus and Sut Jhally. New York: Routledge, 1989. 123–134.

Emerson, Ralph Waldo. "Nature." *The Selected Writings of Ralph Waldo Emerson.* New York: Modern Library, 1950. 3–42.

Fetterly, Judith. *The Resisting Reader: A Feminist Approach to American Fiction.* Bloomington: Indiana University Press, 1977.

Foucault, Michel. *The History of Sexuality: An Introduction.* New York, Vintage, 1980.

———. "Neitzsclie, Genealogy, and History." *Language, Counter-memory, Practice: Selected Essays and Interviews.* Ed. Donald F. Bouchard. Ithaca: Cornell University Press, 1977. 139–164.

Gallop, Jane. "Snatches of Conversation." *Women and Language in Literature and Society.* Eds. Sally McConnell-Ginet, Ruth Borker, and Nelly Furman. New York: Praeger, 1980. 274–283.

Gubar, Susan. "'The Blank Page' and the Issues of Female Creativity." *Writing and Sexual Difference.* Ed. Elizabeth Abel. University of Chicago Press, Chicago: 1982. 243–264.

Haraway, Donna. "Situated Knowledges: The Sciences Question in Feminism and the Privilege of Partial Perspective." *Feminist Studies* 14.3 (Fall 1988): 575–599.

Hawthorne, Nathaniel. *The Scarlet Letter.* Ed. Harry Levin. Boston: Houghton-Mifflin, 1960.

———. "The Birthmark." *Hawthorne's Short Stories.* Ed. Newton Arvin. New York: Vintage, 1955. 147–165.

Herbert, T. Walter, Jr. "Nathaniel Hawthorne, Una Hawthorne, and *The Scarlet Letter:* Interactive Selfhoods and the Cultural Construction of Gender." PMLA 103.3 (May 1988): 285–297.

Johnson, Barbara. "Is Female to Male as Ground is to Figure?" *Feminism and Psychoanalysis*. Eds. Richard Feldstein and Judith Roof. Ithaca: Cornell University Press, 1989. 255–268.

Junichiro, Tanizaki. "The Tattooer." *Seven Tales by Junichiro Tanizaki*. Trans. Howard Hibbett. New York, 1963.

Kafka, Franz. *The Penal Colony*. Trans. Willa and Edwin Muir. New York: Schocken, 1968.

Leverenz, David. "The Politics of Emerson's Man-Making Words." *PMLA* 101.1 (January 1986): 38–56.

Ortner, Sherry. "Is Female to Male as Nature is to Culture?" *Woman, Culture and Society*. Eds. Michelle Zimbalist Rosaldo and Louise Lamphere. Stanford: Stanford University Press, 1974. 67–88.

Sanders, Clinton R. *Customizing the Body: The Art and Culture of Tattooing*. Philadelphia: Temple University Press, 1989.

Scarry, Elaine. *The Body in Pain: The Making and Unmaking of the World*. London: Oxford University Press, 1985.

Wittig, Monique. "The Mark of Gender." *The Poetics of Gender*. Ed. Nancy K. Miller. New York: Columbia University Press, 1986. 63–73.

Contributors

Anne Bolin received her Ph.D. in cultural anthropology from the University of Colorado. Prior to her position as an Assistant Professor of Anthropology at Elon College in North Carolina, she was on the faculty of the University of Colorado at Denver. She has published articles on sex, gender, and women. Her book *In Search of Eve: Transsexual Rites of Passage* (Bergin and Garvey) received the CHOICE Magazine Award for an Outstanding Academic Book for 1988–1989. She is currently co-authoring an anthropology of human sexuality textbook and engaging in ethnographic research with competitive women bodybuilders for a book on women and the body.

Colleen Ballerino Cohen teaches anthropology and women's studies at Vassar College. Her articles and reviews appear in *American Behavioral Scientist, College English, Phoebe, Reviews in Anthropology*, and *Signs*. Her other work on the relationship between gender and narrative style in Ruth Benedict's works, and on postmodern consumer practice, is part of a larger research interest in representation and women's experience. She is currently working on an ethnographic study of identity, nationalism, and tourism in the British West Indies.

Louise Krasniewicz is a visiting assistant professor of anthropology at the University of California, Los Angeles. She is currently researching cultural censorship and the American flag controversy and is interested in the politics of representation and identity in American culture. She is completing a book for Cornell University Press entitled *The Difference Within: The Politics of Representation and the Seneca Women's Peace Encampment*.

Frances E. Mascia-Lees teaches anthropology at Simon's Rock College of Bard where she is also Co-Director of Women's Studies. She is author of *Toward a Model of Women's Status* and editor of a special

171

issue of *Medical Anthropology: Human Sexuality in Biocultural Perspective.* Her writing appears in the *American Anthropologist, American Behavioral Scientist, Medical Anthropology, College English, Signs, Phoebe, American Literary History,* and *Reviews in Anthropology.* She is currently co-authoring a book with Patricia Sharpe which focuses on the representation and display of the subaltern body.

Helena Michie is the author of two books: *The Flesh Made Word: Female Figures, Women's Bodies,* and a volume forthcoming from Oxford University Press on differences among women. She has published articles on feminist theory and Victorian Studies. She is Associate Professor of English at Rice University.

Ñacuñán Sáez teaches Spanish and French at Simon's Rock College of Bard. He has taught at the University of Connecticut and is the recipient of several awards for graduate scholarship. He has participated in numerous professional conferences and has published several articles on literature, history, and semiotics in the United States and South America.

Patricia Sharpe is Co-Director of Women's Studies and member of the faculty in literature at Simon's Rock College of Bard. She has authored book reviews and articles that appeared in *American Literary History, College English, Signs, The Journal of the History of Sexuality, Phoebe, The Michigan Quarterly Review, American Behavioral Scientist, Critical Exchange,* and *Alternative Review.* She is currently at work on a book about feminine subjectivity in the canonical English novel and is co-authoring, with Fran Mascia-Lees, a book entitled *On Display.*

Robyn R. Warhol, Associate Professor of English at the University of Vermont, is author of *Gendered Interventions: Narrative Discourse in the Victorian Novel* (Rutgers University Press, 1989). She has coedited (with Diane Price Herndl) a comprehensive collection of recent contributions to gender studies called *Feminisms: An Anthology of Literary Theory and Criticism,* forthcoming from Rutgers University Press in the Fall of 1991. Her essays in feminist narratology have appeared in *PMLA, Essays in Literature,* and *Psychohistory Review.*